AF834205

International
REVIEW OF
Neurobiology
Volume 69

International
REVIEW OF
Neurobiology
Volume 69

The Neurobiology of C. elegans

EDITED BY

ERIC AAMODT
Department of Biochemistry
Louisiana State University Medical School
Shreveport, Louisiana

ELSEVIER

AMSTERDAM • BOSTON • HEIDELBERG • LONDON
NEW YORK • OXFORD • PARIS • SAN DIEGO
SAN FRANCISCO • SINGAPORE • SYDNEY • TOKYO
Academic Press is an imprint of Elsevier

Elsevier Academic Press
525 B Street, Suite 1900, San Diego, California 92101-4495, USA
84 Theobald's Road, London WC1X 8RR, UK

This book is printed on acid-free paper. ♾

For all information on all Elsevier Academic Press publications visit our Web site at www.books.elsevier.com

ISBN-13: 978-0-12-366870-7
ISBN-10: 0-12-366870-0

Printed and bound in the United Kingdom
Transferred to Digital Printing, 2011

CONTENTS

Sexual Behavior of the *Caenorhabditis elegans* Male

SCOTT W. EMMONS

The Motor Circuit

STEPHEN E. VON STETINA, MILLET TREININ, AND DAVID M. MILLER, III

Mechanosensation in *Caenorhabditis elegans*

ROBERT O'HAGAN AND MARTIN CHALFIE

CONTRIBUTORS

Numbers in parentheses indicate the pages on which the authors' contributions begin.

Eric Aamodt (73), Department of Biochemistry, Louisiana State University Health Sciences Center-Shreveport, Shreveport, Louisiana

Stephanie Aamodt (73), Department of Biological Sciences, Louisiana State University in Shreveport, Shreveport, Louisiana

Zeynep Altun (1), Center for *C. elegans* Anatomy, Department of Neuroscience, Albert Einstein College of Medicine, Bronx, New York; Department of Psychiatry and Behavioral Sciences, Montefiore Medical Center, Bronx, New York

Martin Chalfie (169), Department of Biological Sciences, Columbia University, New York, New York

Scott W. Emmons (99), Department of Molecular Genetics, Albert Einstein College of Medicine, Bronx, New York

Andrew C. Giles (37), Department of Psychology and Brain Research Centre, University of British Columbia, Vancouver, British Columbia, Canada

David H. Hall (1), Center for *C. elegans* Anatomy, Department of Neuroscience, Albert Einstein College of Medicine, Bronx, New York

Robyn Lints* (1), Center for *C. elegans* Anatomy, Department of Neuroscience, Albert Einstein College of Medicine, Bronx, New York

David M. Miller, III (125), Department of Cell and Developmental Biology, Vanderbilt University, Nashville, Tennessee

Robert O'Hagan[†] (169), Department of Biological Sciences, Columbia University, New York, New York

Catharine H. Rankin (37), Department of Psychology and Brain Research Centre, University of British Columbia, Vancouver, British Columbia, Canada

Jacqueline K. Rose (37), Department of Psychology and Brain Research Centre, University of British Columbia, Vancouver, British Columbia, Canada

Millet Treinin (125), Department of Physiology, Hebrew University-Hadassah Medical School, Jerusalem, Israel

Stephen E. Von Stetina (125), Department of Cell and Developmental Biology, Vanderbilt University, Nashville, Tennessee

*Current address: Department of Biology, Texas A & M University, College Station, Texas.

[†]Current address: Massachusetts Institute of Technology, 77 Massachusetts Avenue, Cambridge Massachusetts.

PREFACE

In 1962 at the MRC in Cambridge England, Sydney Brenner and Francis Crick, who shared an office, began discussing what big questions should be addressed next in molecular biology. They concluded that the next two big challenges were to understand how development occurs and how nervous systems work.

To address these problems, they decided that the best system would be a small metazoan with a short life cycle that was amenable to genetics and could be easily grown and handled in large numbers. Their work with phage and bacteria had shown them the power of a focused approach on a single simple-model organism. They wanted something with few cells where the complete cell lineage could be determined and the anatomy could be described in electron microscopic detail.

Dr. Brenner then went to the library to look for such a beast and eventually settled on *Caenorhabditis elegans* as the best organism for addressing fundamental questions in neurobiology and development. *C. elegans* is a microscopic ($\sim$1 mm) nematode (roundworm) that normally lives in soil, but can be easily grown in the laboratory.

Working together with a small group of extremely talented postdoctoral fellows, Sidney Brenner set out to tame this small metazoan. He began by isolating and mapping visible mutations and organizing them into a genetic map (Brenner, 1974). At the same time, his assistant, Nichol Thomson, began to make electron micrographs of serial sections along the entire length of the worm and John White began mapping the nervous system. These studies eventually led to John White and colleagues' epic monograph describing the structure of the entire *C. elegans* nervous system down to the position of each synapse (White *et al.*, 1986). Brenner's colleagues, John Sulston and Bob Horvitz, determined the larval cell lineage of the *C. elegans* hermaphrodite (Sulston and Horvitz, 1977) and John Sulston determined the embryonic cell lineage (Sulston *et al.*, 1983).

Since theses early beginnings, the *C. elegans* field has grown exponentially until today; there are thousands of researchers working with this simple animal and *C. elegans* stands as the best-understood animal in the world in terms of anatomy, genetics, development, and behavior. In 2002 Sydney Brenner shared the Nobel prize with John Sulston and Robert Horvitz for the studies described above and much more.

C. elegans has many advantages as an experimental organism. It can be easily grown on Petri plates containing bacteria, which serves as food for these animals and it can be grown en masse in liquid culture. Its life cycle takes about three days under typical laboratory conditions. There are two sexes: hermaphrodites and males. The hermaphrodites produce sperm early in development and can use this sperm to fertilize their own eggs. The males can mate with the hermaphrodites and the male sperm is used preferentially over the hermaphrodite sperm. Hermaphrodites cannot fertilize each other. An unfertilized hermaphrodite can lay about 300 eggs and a fertilized hermaphrodite can lay over a thousand eggs. The hermaphrodite genetics makes it easy to homozygose and maintain recessive mutations, and the males allow for genetic crosses.

C. elegans is very simple animal both genetically and morphologically. The adult hermaphrodite has 959 somatic nuclei and just 302 neurons (Wood, 1988). *C. elegans* was the first metazoan to have its genome sequenced (Consortium, 1998). The genome contains just 100,258,171 base pairs and 19,100 predicted protein-encoding genes in five pairs of autosomes, and a pair of X chromosomes. It is easy to transform *C. elegans* with exogenous DNA. *C. elegans* is particularly susceptible to RNAi, and simply feeding animals bacteria that are expressing a specific double stranded RNA can neutralize many of its genes.

C. elegans are transparent at each stage of their life cycle so the animal can be analyzed at the cellular level by light microscopy, or in animals that are expressing transgenes that encode fluorescent proteins, by fluorescence microscopy.

In addition to the genome sequence, other studies have provided information about many of the genes. Expressed sequence tags have confirmed the expression and splicing of mRNAs. The mRNA expression levels have been analyzed by microarray and SAGE analysis. The loss of function phenotypes of many genes have been determined by massive RNAi studies and the *C. elegans* Gene Knockout Consortium is systematically knocking out every gene in *C. elegans* by isolating site directed deletions. The expression patterns of numerous genes are being determined by *in situ* hybridization and GFP gene fusions. Massive yeast two-hybrid screens and pull-down experiments are testing the interaction of each protein domain with every other. All of this data is easily accessible through an on-line database, WormBase, which is located at http://www.wormbase.org/.

This volume assembles together a series of chapters that describe the advances this system has made toward solving some of the major problems in neurobiology. The first chapter provides an introduction to the anatomy of the *C. elegans* nervous system, which will help the reader understand each of the chapters that follow. The next two chapters on learning, conditioning and memory, and neuronal specification and differentiation, summarize the current state of the *C. elegans* field in these two major areas of neurobiology. Each of the chapters after that describes studies in *C. elegans* that have provided particularly exciting insights into neurobiology.

Although it has been 43 years since Sidney Brenner and Francis Crick began their discussions that eventually led to our current understanding of the *C. elegans* nervous system, it is clear from the number of exciting papers published every week on some new aspect of *C. elegans* neurobiology that we still have much to learn with this powerful model system.

Eric Aamodt

References

Brenner, S. (1974). The genetics of *Caenorhabditis elegans*. *Genetics* **77,** 71–94.

Consortium, T. C. e. G. S. (1998). Genome sequence of the nematode *Caenorhabditis elegans*: A platform for investigating biology. *Science* **282,** 2012–2018.

Sulston, J. E., and Horvitz, H. R. (1977). Post-embryonic cell lineages of the nematode *Caenorhabditis elegans*. *Dev. Biol.* **56,** 110–156.

Sulston, J. E., Schierenberg, E., White, J. G., and Thomson, J. N. (1983). The embryonic cell lineage of the nematode *Caenorhabditis elegans*. *Dev. Biol.* **100,** 64–119.

White, J. G., Southgate, E., Thomson, J. N., and Brenner, S. (1986). The structure of the nervous system of *Caenorhabditis elegans*. *Philos. Trans. R. Soc. Lond. B Biol. Sci.* **314,** 1–340.

Wood, W. B. (1988). The Nematode *Caenorhabditis elegans*. Cold Spring Harbor Press, Cold Spring Harbor.

NEMATODE NEURONS: ANATOMY AND ANATOMICAL METHODS IN *CAENORHABDITIS ELEGANS*

David H. Hall,* Robyn Lints,*,[1] and Zeynep Altun*,†

*Center for *C. elegans* Anatomy
Department of Neuroscience
Albert Einstein College of Medicine
Bronx, New York
†Department of Psychiatry and Behavioral Sciences
Montefiore Medical Center, Bronx, New York

The nervous system of *Caenorhabditis elegans* consists of 358 cells (473 in males) including sensory, motor, interneurons, polymodal neurons, and related glial cells that comprise 118 subgroups according to their body position, synaptic patterns, and sensory specializations. Of the 302 neurons of the adult hermaphrodite, the 20 pharyngeal neurons function virtually independently of the 282 somatic neurons of the bodywall. Three classic papers have described the nervous system in great detail for both genders based upon electron microscopy (EM) and serial section reconstructions (Albertson and Thomson, 1976; Sulston *et al.*, 1980;

[1]Current address: Department of Biology, Texas A & M University, College Station, Texas.

DOI: 10.1016/S0074-7742(05)69001-0

0074-7742/06 $35.00

White *et al.*, 1986). Neuronal cell fates are reproducible from animal to animal, and are determined mostly by lineage rather than their neighborhoods, just as for other cell fates in *C. elegans*. Most neurons in the nematode are extremely simple in shape and organization compared to higher animals. Yet the following chapters in this volume describe in detail how these simple cells are still capable of performing complex tasks during the animal's range of normal behaviors. Neuronal sensory specializations most often consist of a unitary ciliated process, often wrapped by one or two glial cells at the receptive ending. Synaptic connections rarely involve separate branches or boutons, but instead occur *en passant* at local swellings of the primary neurite. Most neurons grow few or no branches, unlike the neuronal arborizations in higher organisms. Neuronal synaptic zones (the active zone at a synapse) are also small and simple – so much so that they are easily overlooked except at the electron microscopy level. In this chapter, we will review a few of the neuron subgroups whose shapes and properties are of current interest, and discuss anatomical techniques useful for exploring and defining neurons in *C. elegans*.

I. Introduction

The nematode nervous system is less centralized than those in higher animals (Bullock and Horridge, 1965), although the nerve ring in the head does bring together the majority of the body's neurons and synapses. The adult male also has enlarged tail ganglia (gaining 89 extra neurons) that function as a major processing center in the tail to control aspects of male mating behavior (discussed in Chapter 4). The largest proportion of synaptic neuropil is consolidated in the nerve ring itself. There is no separate neuropil associated with most clusters (ganglia) of cell bodies that surround the ring (Table I). Aside from the nerve ring and its extension beneath the nearby retrovesicular ganglion, the other principal sites of synaptic interactions consist of the ventral and dorsal nerve cords (discussed in Chapter 5) and in the preanal ganglion in the tail (Hall and Russell, 1991). A limited synaptic zone is associated with the egg-laying system at the vulva (Thomas *et al.*, 1990). Synapses in other peripheral locales are rare. The nematode was the first eukaryote in which all neuronal and glial cell fates could be identified and characterized. The comprehensive nature of our knowledge of physical, molecular, and developmental aspects of these cells potentiates *C. elegans* as a model system for exploring both the development and function of complete behavioral circuits as well as the genetic specification of neuronal characters in isolation. Similarities between nematode neurons and those in other organisms include the early guidance receptors used for axon outgrowth, and the cellular machinery (e.g., vesicular transporters, ion channels, synaptic proteins) utilized to carry out mature functions (Ruvkun and Hobert, 1998) (see Chapter 3). The

TABLE I
SOMATIC GANGLIA AND THEIR ATTRIBUTES

Name	Cell types[a]	Neuropil[e]	Locale	Cell count[a]
Anterior ganglion and	SN IN MN	No	Front nerve ring	17/17(19/19)
its ant. subgroup	GL	No	Near procorpus	7/7
Dorsal ganglion	MN PM	No	Near nerve ring	6
Lateral ganglia	SN IN MN GL	No	Near nerve ring	32/33
Ventral ganglion	MN IN	Yes	Behind nerve ring	32
Retrovesicular ganglion	MN IN	Yes	Ant. to ventral cord	18
and nearby lateral cells	SN MN GL	No	Beside second bulb	6/7
Ventral cord[b]	MN	Yes[d]	Midbody	62 (72)
Posterior lateral ganglia	SN GL IN	No	Posterior midbody	6/4
Pre-anal ganglion	MN IN (SN GL)	Yes	Behind ventral cord	12 (28)
Dorso-rectal ganglion	MN IN	No[c]	Tail	3 (9)
Lumbar ganglia	SN GL IN PM	No[c]	Tail	14/14 (50/50)
Cloacal ganglia (male only)	SN IN PM GL	Yes	Tail	0 (7/7)

[a]L/R sides, including glia; numbers or cell types in parentheses are for male ganglia.

[b] Technically not a ganglion, but contains the largest set of neurons and synapses. Cell count for the ventral cord does not include those MNs that lie in the retrovesicular ganglion (DA1, VA1, etc) or pre-anal ganglion (DA8, DA9, VA12, and so on).

[c]Yes in male.

[d]Nerve ring and dorsal cord are the other major regions of synaptic neuropil.

SN = sensory neuron; IN = interneuron; MN = motor neuron; PM = polymodal neuron; GL = glial cell. Some cells do not associate with a ganglion, but lie free in the pseudocoelom, or alone along the hypodermis and are not counted above. Includes GLRs, HSNs, URXs, BDUs, ALMs, CANs, and SDQR.

[e]Indicates a local synaptic zone associated with the ganglion.

ligands used for cell signaling and the transmitters (both classical neurotransmitters [such as GABA, acetylcholine, serotonin, and so on] and neuropeptides) used in neuronal signaling are generally conserved as well (Bargmann, 1998) (see Chapter 3).

Classification of nematode neurons into functional categories was initially done according to transmission electron microscopy (TEM) data alone (Albertson and Thomson, 1976; Hall, 1977; Sulston *et al.*, 1980; Ward *et al.*, 1975; Ware *et al.*, 1975; White *et al.*, 1976, 1986). Motor neurons comprised all of those cells that consistently formed neuromuscular junctions (NMJs) onto one or more muscle cells. Sensory neurons were those cells that possessed a distinct sensory dendrite of any modality (although the specific modality often awaited physiological testing, laser ablation, or mutant behavior). Interneurons comprised most of the rest; where their synaptic wiring imputed some role in collecting and relaying synaptic activity from upstream (sensory or interneuron) neurons to downstream effector neurons. Some neurons that have been categorized as "sensory" or "motor" cells may also have a high degree of interconnectivity that would also classify them as interneurons.

Some neurons displayed dual roles ("polymodal"), for instance having both a sensory dendrite and NMJ outputs. Several neurons with few distinctive features were difficult to categorize. Anatomic and genetic studies have identified several "pioneer neurons" that play important roles in axon guidance or maintenance, while having little or no role in synaptic circuits (Aurelio *et al.*, 2002; Chen *et al.*, 2005; Durbin, 1987; Wadsworth *et al.*, 1996). Some potential neurons undergo early apoptotic deaths soon after their birth and could not be categorized into a functional class (Sulston *et al.*, 1983). Some neurons change their principal wiring features during development, adding new synapses and deleting others (Walthall, 1990; White *et al.*, 1978). Evidence is less solid regarding whether nematode neurons can modify their circuitry in response to the animal's experience, although their patterns of activity must surely vary according to this experience (Chao *et al.*, 2004; Gomez *et al.*, 2001; Peckol *et al.*, 1999, 2001; Zhao and Nonet, 2000) (see Chapter 2). Regions important for sensory transduction in a nematode neuron are not always identifiable by their TEM structure alone, and thus certain cells evade easy categorization. For these reasons, the final tabulation of neuron subclasses has required occasional corrections as more physiological and developmental data has accumulated.

The basic structure of a *C. elegans* neuron consists of a monopolar or bipolar cell extending one or two simple thin processes (neurites) along the length of the body (Fig. 1). The cell body is generally about 2 μm in diameter and contains a small compact nucleus. The neurites are extremely narrow in diameter (often just $\leq$0.2 μm) and most have virtually no local branches. Since the typical neuron is essentially unbranched, it must send its neurite along a local "neighborhood" in close proximity with its chosen partners before forming many synapses (White *et al.*, 1983). Synapses occur *en passant* at local swellings along the length of a neurite (Fig. 2). A neurite often ends by making a gap junction (electrical synapse) where it contacts the neurite of a homologous neuron, such that neither cell extends a process into a territory occupied by its homologues (thus, the axon of VD4 will form junctions at its anterior and posterior limits with the axons of VD3 and VD5 in the ventral cord, and axons of left/right neuron pairs often form junctions where they meet in the nerve ring). Thus, the nervous system is exceedingly economical not only in cell number but also in the number of processes or branches. Even the final positions of neuron cell bodies seem efficiently managed to minimize the total "wiring cost" of the nervous system (Chen *et al.*, 2005). The total number of synapses per neuron is also remarkably limited. The hermaphrodite adult contains about 5000 chemical synapses, 700 gap junctions and 2000 NMJs. Many cells only form 25 to 50 synapses, and certain neurons form fewer than 10 synapses in total, including repeated contacts to the same partner. Even the most important command interneurons have about 200 synapses in total.

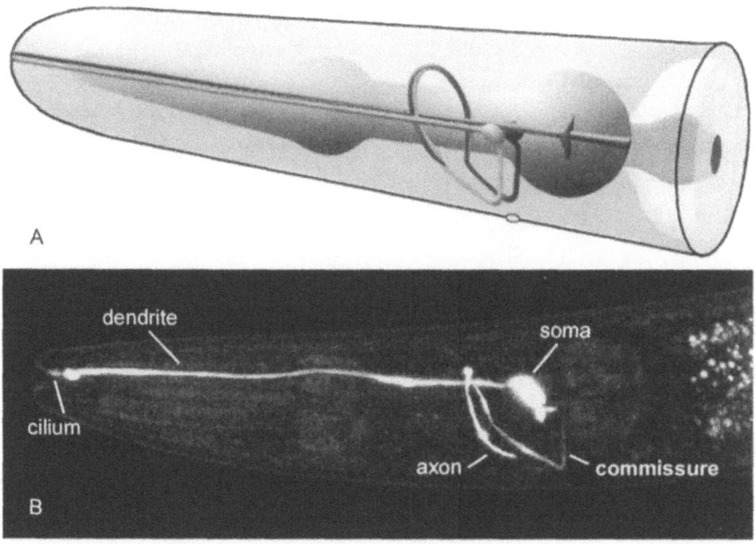

FIG. 1. ASE neuron morphology. (A) Cartoon shows ASEL (light cell) and ASER (dark cell) and their processes in relation to the pharynx. Each sensory neuron sends one dendrite anteriorly to the nose, and a single axon into the nerve ring, circling around the isthmus between the first and second bulbs of the pharynx. Neither the dendrite nor the axon has any local branching, but their trajectories closely match those of their bilateral homologues. (B) Whole mount *gcy-5*:GFP construct of a living adult in which the ASER neuron is specifically labeled, including the soma, dendrite, and axon, as well as the ventral commissure by which the axon reaches the nerve ring. The dendrite is moderately larger at the base of the cilium, and the axon has several local swellings within the nerve ring neuropil. Micrograph is courtesy of David Miller, Vanderbilt University. ~5000X.

II. Categorization of Neuron Cell Fates

A. Sensory Neuron Classes

Many sensory cells have specialized dendritic processes that end up embedded into the bodywall cuticle or exposed to the outside environment via holes in the cuticle. These cells may be activated by physical deformation of the cuticle or chemical stimuli that reach a ciliated dendrite via the opening (e.g., neurons involved in chemotaxis and odortaxis behavior, or nose-touch response) (Bargmann, 1993; Bargmann and Kaplan, 1998; Troemel *et al.*, 1995). A few other sensory processes are linked to movable parts of the worm anatomy (such as the male spicules, the extreme tail tip, or the inner surface of the pharynx); these cells are suggested to function in proprioception, sensing deep touch, or sensing the presence of food inside the pharyngeal lumen. Volatile odorants can reach

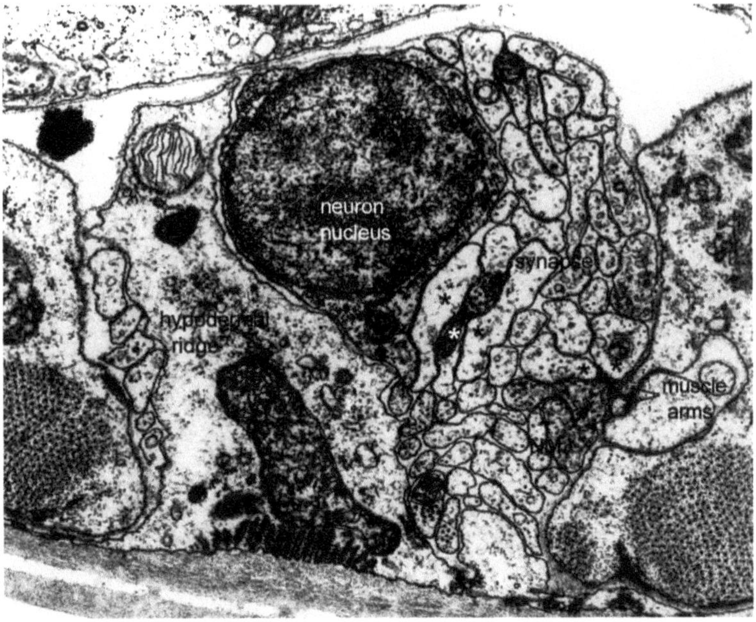

FIG. 2. Synapse morphology. Two arrows indicate the similar structural features of a triadic chemical synapse and a NMJ. The presynaptic axons for each contact display a collection of synapse vesicles next to a presynaptic density on the cell membrane. Asterisks mark the several nearby neuronal processes that are expected to be postsynaptic at each release zone. Two small muscle arms (thin processes) lie opposed to the release zone at the NMJ and are believed to permit simultaneous stimulation of two bodywall muscle cells from this contact. Contractile elements of these muscles are evident to the left and right sides of the nerve cord. Thin section of wild-type adult ventral nerve cord from work of John White (MRC/LMB archive, now housed at the Center for *C. elegans* Anatomy in New York), ~50,000X.

sensors embedded deeper in the amphid sheath that are not exposed directly to the outside (AWA, AWB, AWC) (Sagasti *et al.*, 1999). Some receptor processes lie deep inside the body (AQR, PQR, and URX), but can still sense environmental stimuli such as ambient oxygen levels (Gray *et al.*, 2004). Several sets of mechanosensory neurons (ALM, PLM) extend microtubule-filled processes along the cuticle that respond to "light touch" onto the cuticle, acting via protein links to the underlying basal lamina (Chalfie *et al.*, 1985; Emtage *et al.*, 2004) (see Chapter 6). Some neurons extend long unspecialized processes along the body that have been suggested to act as stretch receptors (proprioceptors; PHC, VA, VB, PCB, ALN, PLN) (Hall and Russell, 1991; Sulston *et al.*, 1980; White *et al.*, 1986). *C. elegans* have been reported to have a photo response, but no sensory receptor is yet known to be responsible (Burr, 1985). Genetic, physiological and ablation studies have proven crucial in exploring the range of sensory capabilities of these various cell types. Such investigations are still expected to uncover the sensory capabilities of additional cryptic sensory cells. Table II provides a current list for the known sensory neurons for both genders. Although not detailed in Table II, there is growing evidence that some of these cells respond to more than one modality. For example, the ASH sensors of the amphid respond well to many volatile and soluble chemorepellants, to high osmotic strength, and to mechanical touch (Hilliard *et al.*, 2004; Tobin and Bargmann, 2004).

A few sensory neurons adopt much more complex arbors than most other neurons in the nematode. Some of these more highly branched neurons were well described in serial thin section reconstructions. For instance, the AFD neuron's distal dendrite ends in a profusion of about 50 thin branches that look much like villi (Cassata *et al.*, 2000; Ward *et al.*, 1975; Ware *et al.*, 1975). This specialized dendrite seems to be responsible for transducing temperature (Hedgecock and Russell, 1975; Mori and Ohshima, 1995). The PVD neuron has a web of fine branches extending along the bodywall, which were noted first in live animals expressing several gfp constructs in PVD (Halevi *et al.*, 2002; Tsalik *et al.*, 2003). The function of these branches is still uncertain, but they may provide a network of sensory endings over the whole midbody region, important for nociception (Kaplan and Horvitz, 1993; Suzuki *et al.*, 2003; Tobin and Bargmann, 2004). The FLP sensory neurons have just a few extra branches along their dendrite; their function is currently unknown.

The male ray sensory neurons have rather branchy axons reaching into the preanal ganglion from their cell bodies in the lumbar ganglia. The preanal ganglion is their principal region for finding synaptic partners, and each ray neuron may form 5 to 10 branches there (Xu, Albertson, Thomson, Hall and Emmons, 2005, unpublished observations). The individual rays sometimes differ substantially in the portions of the pre-anal ganglion (PAG) that they invade, and their individual branches can form chemical synapses with widely different partners on separate branches.

B. Motor Neuron Classes

There are distinct classes of motor neurons specialized to operate different portions of the nematode anatomy.

1. *Control of Head, Neck, and Bodywall Muscles*

Control of head muscles is accomplished by motor neurons whose NMJs lie in the nerve ring. Bodywall muscles are primarily controlled by a column of cells lying along the length of the body in the ventral nerve cord (see Chapter 5). Control of the neck muscles is shared by motor neurons acting in the nerve ring and the ventral cord cells (White *et al.*, 1976, 1986). The ventral cord motor

TABLE II

SENSORY SYSTEMS IN *C. ELEGANS* (AFTER WHITE *ET AL.*, 1986)

Chemosensors (exposed cilia via holes in cuticle)
(Hall and Russell, 1991; Hilliard *et al.*, 2002, 2004; Perkins *et al.*, 1986; Sulston *et al.*, 1980; Ward *et al.*, 1975; Ware *et al.*, 1975; White *et al.*, 1986)

- Amphids
  - ADF L/R
  - ADL L/R
  - ASE L/R
  - ASG L/R
  - ASH L/R
  - ASI L/R
  - ASJ L/R
  - ASK L/R
- Inner labials
  - IL2 (6 lips)
- Phasmids
  - PHA L/R
  - PHB L/R
- Cephalic
  - CEM (4 lips) (male only)
- Rays (male only)
  - RnB L/R (8 "open" rays)
- Hook (male only)
  - HOB
- Spicules (male only)
  - SPD L/R
  - SPV L/R

Odorant sensors (embedded cilia in lips)
(Sagasti *et al.*, 1999; Troemel *et al.*, 1995, 1997; Ward *et al.*, 1975; Ware *et al.*, 1975)

- Amphids
  - AWA L/R
  - AWB L/R
  - AWC L/R

Thermosensors (embedded cilia in lips)
(Mori and Ohshima, 1995; Ward *et al.*, 1975; Ware *et al.*, 1975)

- Amphids
  - AFD L/R

Mechanosensors (unexposed cilia ending in cuticle)
(Hall and Russell, 1991; Perkins *et al.*, 1986; Sawin *et al.*, 2000; Sulston *et al.*, 1980; White *et al.*, 1986)

- Inner labials
  - IL1 (6 lips) (cilium does not protrude through the hole)
- Cephalic
  - CEP (4 lips)
- Outer labials
  - OLQ (4 lips)
  - OLL L/R

TABLE II (*Continued*)

Deirids and postdeirids
ADE L/R
PDE L/R
Hook (male only)
HOA (cilium does not protrude through the hole)
Post-cloacal sensillum (male only)
PCA L/R (polymodal MNs, protruding cilium, but in cuticle)
PCB L/R (polymodal MNs)
PCC L/R (polymodal MNs, striated rootlet, no cilium)
Tail tip
PVR
Rays (male only)
RnA L/R (9 rays)
R6B L/R (ray 6, which is closed)
Mechanosensors (microtubule-filled dendrites)
(Chalfie *et al.*, 1985)
ALM L/R
AVM (minor role in mechanosensation)
PLM L/R
PVM (minor role in mechanosensation)
Oxygen sensors (ciliated dendrite in pseudocoelom or tail)
(Cheung *et al.*, 2004; Gray *et al.*, 2004; White *et al.*, 1986)
AQR
PQR
URX L/R
Oxygen sensors (unexposed cilium in the lips)
(Gray and Bargmann, 2005, personal communication; Ward *et al.*, 1975; Ware *et al.*, 1975)
BAG L/R
Proprioceptors (having putative stretch-sensitive process)
(Albertson and Thomson, 1976; Hall, 1977; Hall and Russell, 1991; Tobin and Bargmann, 2004; White *et al.*, 1986)
Bodywall
ALN L/R
PLN L/R
VA (12 cells) (polymodal MNs)
VB (11 cells) (polymodal MNs)
PVD L/R (nociception)
Tail tip
AVG
PDB
PHC L/R
Spicules (male only)
SPC L/R (polymodal MNs)
Pharynx
I1 L/R (polymodal interneurons)
I2 L/R (polymodal interneurons)
I3 (polymodal interneuron)
I5 (polymodal interneuron)

(*Continued*)

TABLE II (*Continued*)

I6 (polymodal interneuron)
MC L/R
M3 L/R (polymodal MNs)
NSM L/R (polymodal secretory neurons)
Unknown modality (buried endings)
(Kaplan and Horvitz, 1993; White *et al.*, 1986)
AUA L/R (polymodal interneurons)
FLP L/R (nose touch)
URY L/R
URA (4 cells) (polymodal MNs)
URB L/R (polymodal interneurons)

n.b. Many of the above functions are provisional, based upon the structure and locale of the sensory receptors. Neurons and their sensors are assigned to only one category here, although they may respond to many cues *in vivo*.

neurons fall into three distinct anatomical classes (excitatory classes A and B, inhibitory class D) and are thought to drive anterior and posterior body motion. The dorsal and ventral muscle rows are innervated by these neurons in a segmental, non-overlapping fashion. In contrast to most organisms, *C. elegans* muscles receive neuronal input by sending arms to the major nerve cords, rather than neurons sending branches to the target muscles.

2. *Control of Pharyngeal Pumping*

The nervous system of the pharynx has five classes of motor neurons (M1 to M5), all intrinsic to the pharynx and functioning as an autonomous unit. (Albertson and Thomson, 1976; Avery and Horvitz, 1989; Avery and Thomas, 1997). The motor neurons form their NMJs in a series of contacts lying deep inside the pharyngeal muscles. Some of these motor contacts occur at or near the two pharyngeal "nerve rings" of the first and second bulb, while others occur at the extreme anterior end of the pharynx. In between, the pharyngeal neuron processes travel lengthwise in narrow channels lying between rows of muscle cells. Unlike body wall muscles, pharyngeal muscles do not extend any arms to the motor neurons.

Since the neurons of the pharynx lie in a rather unique locale, their exact shapes are somewhat different from those in the somatic nervous system, but overall they show the same simplicity, lack of branching, *en passant* synapses, and relatively few synaptic contacts per cell.

3. *Control of Male Tail-Mating Muscles*

The male tail includes 41 specialized muscles required for various aspects of male-mating behavior (perhaps the most complex behavioral actions controlled by the nematode nervous system) (Emmons and Sternberg, 1997) (see Chapter 4). The complete wiring diagram for those muscles is still not determined (Sulston *et al.*, 1980; Xu *et al.*, 2005, unpublished observations). However, the male has a large number of gender-specific neurons lying in the tail region that are the principal players in driving these behaviors.

Curiously, there are rather few pure gender-specific motor cells in the tail ganglia for driving these muscles. Instead, it is notable that many of the male sensory neurons synapse directly onto certain muscles. These "polymodal" sensorimotor cells include a subset of the 18 ray neurons (RnA), and neurons of the spicules and the postcloacal sensilla (Lints *et al.*, 2005, unpublished observations; Sulston *et al.*, 1980). In a few instances these sensorimotor neurons have synaptic branches that reach directly to specialized muscle cells, rather than having a muscle arm reach out to the presynaptic axon. We will discuss several examples of polymodal neurons in the following text.

4. *Control of Egg-Laying*

Two classes of motor neurons principally innervate the vulval muscles, HSN and VC (Chalfie and White, 1988; Thomas *et al.*, 1990; White *et al.*, 1986). Unlike most motor neurons, these cells actually form many short branches that invade a region near the vulval opening to reach their target muscles, the vm2 cells. Other vulval and uterine muscles are linked to vm2 via gap junctions (White *et al.*, 1986). One of the vm1 muscles also receives a few direct chemical synapses from nearby motor neurons of the ventral cord, including VA7, VB6, and VD7. The many gap junctions occurring between the assorted muscle classes suggest that vulval opening and constriction of the uterine egg chamber may be coordinated in timing and power.

The HSN and VC neurons are some of the more highly branched motor cells in the nematode (White *et al.*, 1986). HSN branches seem to be restricted to the local neuropil near their target muscles. However, VC axons also make occasional short branches all along the course of their ventral cord trajectories. The VC neurons make scattered contacts to many ventral bodywall muscles, whereas the HSNs are purely involved in contacts to the vulval muscles.

5. *Control of Defecation*

There are several distinctive enteric muscle cells that act coordinately in defecation, and these muscles contract in a distinctive rhythmic pattern that takes almost one minute to occur (Avery and Thomas, 1997; Liu and Thomas, 1994). A calcium signal originating in the posterior intestine is suggested to be crucial in initiating each defecation cycle (Dal Santo *et al.*, 1999) as well as activating the two motor neurons AVL and DVB which drive contraction of the posterior body (Avery and Thomas, 1997; McIntire *et al.*, 1993). Thin muscle arms extend from all of these defecation muscles to reach the top of the preanal ganglion, where the DVB axon forms a series of vesicle-filled swellings (Hall and Russell, 1991; White *et al.*, 1986).

C. Interneuron Classes

Interneurons act as intermediaries between other neurons, receiving inputs from one or more classes, processing that information and relaying outputs to follower cells.

The definition of an interneuron is somewhat troubling in this animal, since most neurons have relatively few synapses in total, and some nominal interneurons seem to have quite few synapses at all. While a few prominent interneurons are easy to recognize, such as the "command interneurons" of the ventral nerve cord (see the following text), there are some cells that seem to fall into the interneuron category by default. Thus any neuron lacking an obvious sensory cilium and lacking obvious neuromuscular output may obtain this classification, irrespective of its abundant or negligible importance in the wiring. Some may be better described as pioneer neurons or developmental neurons (see the following text).

A prominent group in this class are the command interneurons for locomotion (see Chapter 5). These include the AVA, AVB, AVD, and PVC interneurons that make repeated contacts to all the motor neurons in the ventral cord, while collecting abundant synaptic inputs in the nerve ring and in the tail ganglia from many sense cells (White *et al.*, 1976, 1986). These interneurons influence the decision of the animal to reverse direction or to maintain its current body motion according to a cascade of sensory inputs and behavioral imperatives.

Certain interneurons in the nerve ring have synaptic inputs from a variety of sensory neurons and may use these multiple inputs to select which environmental signals are more important to attend, especially related to the presence or absence of food. A well-fed animal may attend closely to pheromones from the opposite gender, while a starving animal will be more interested in finding a new food source. Signals from the many chemosensory neurons of the amphids in the head impinge first onto the AIA and AIB interneurons (White *et al.*, 1986). It is assumed that they integrate these many inputs; curiously AIA synapses most heavily onto AIB, and AIB makes limited outputs to several more classes of interneurons. Other interneurons seem to integrate sensory inputs from two opposing classes of sensors to help in discrimination of important cues. For instance, temperature signals received by the AFD sense cells are compared in AIY and AIZ interneurons with an unknown secondary signal (from a cryptic sense cell) to detect significant changes over time (e.g., detecting a temperature gradient). These cells relay their outputs to RIA interneurons and perhaps to others to decide whether the animal should move towards the warmer or colder portion of the environment (Mori, 1999; Mori and Ohshima, 1995). Laser ablation of either AIY or AIZ interneurons restricts the ability of the animal to respond normally; although the sensors remain intact, the animal remains motile, and it still reacts to a temperature gradient. The ablated animals tend to make poor choices compared to unablated ones, by moving up or down the temperature gradient to very hot or cold regions.

D. "Sensorimotor" or "Polymodal" Neurons

The small size and low total cell number in *C. elegans* forces the animal to be rather economical in deploying its neurons to various tasks. Many neurons perform both sensory and motor roles, or perform both as major interneurons while also having a sensory or motor function, as we briefly review in the following text.

In the pharynx, many neurons may be polymodal, since most interneurons have processes that are tightly bound to the inner cuticle, suggesting a secondary role in proprioception (Albertson and Thomson, 1976). The NSM neurons of the pharynx are more clearly polymodal, as they function both as motor neurons and as neurosecretory cells, releasing an unknown factor into the pseudocoelom. Similarly, the MI neuron of the pharynx is both an important motor neuron and an active interneuron (Albertson and Thomson, 1976).

In the head, the IL1 neurons also seem to be polymodal (White *et al.*, 1986). They likely have mechanoreceptors in the lips and their axons synapse directly onto the head muscles at the muscle plate beneath the nerve ring. Other neurons that may fit this dual role could include the RIV and AVL neurons of the nerve ring, which combine motor and interneuron functions (White *et al.*, 1986).

As previously mentioned, neurons sharing both motor and sensory functions are also found in the male nervous system. For instance, many of the male-specific sensory neurons of the rays also have direct NMJs to local male-specific muscles. Several postcloacal sensilla, including PCA, PCB, and PCC neurons have dual roles as sense cells and as motor neurons acting on the gubernacular and oblique muscles of the tail (Lints *et al.*, 2005, unpublished results; Sulston *et al.*, 1980; also see http://worms. aecom.yu.edu). Another example is the pair of spicule-associated SPC neurons. Each SPC neuron has separate sensory processes attached to the spicule's dorsal protractor muscle and to the anal depressor muscle, and has an axon synapsing directly onto the dorsal and ventral spicule protractor muscles. Again, more details are shown at http://worms.aecom.yu.edu.

Polymodal neurons with important roles in axon guidance as well as roles in synaptic wiring might include AVG, PVN, and PLN, which are all discussed in the following text.

E. Glial Cells

There are a number of prominent cell types that seem to adopt glial cell fates, either shielding the distal endings of ciliated neurons or creating local compartments that help to isolate/insulate certain synaptic regions. Examples of the former category would be the numerous sheath and socket cells that wrap the sense endings of the lips, amphids, and phasmids, and the structural cells that wrap the male ray tips (Perkins *et al.*, 1986; Sulston *et al.*, 1980). We argue in the following that the GLR cells may not be acting as glial cells per se, but there are valid counterarguments. The CEP sheath cells are examples of the second category. They form glial-like wrapping of portions of the nerve ring neuropil and the ventral ganglion neuropil (Altun *et al.*, 2005; see http://www.wormatlas.org/handbook/hypodermis/CEPimage%20gallery.htm). Although the CEP wrappings are incomplete here, it seems possible that they are rudimentary homologues to the more elaborate glial wrappings of synaptic glomeruli in higher animals (Bullock and Horridge, 1965; Shepard, 1972).

III. What Defines a Neuron? Synapses, Scaffolds, and Pioneers

There are a few odd cells known in the *C. elegans* anatomy that look vaguely like neurons, or that have sometimes been included as neurons in comprehensive cell listings, but that remain difficult to classify. Most of these cells are embedded in the bodywall epithelium; some of these oddballs may be acting as gland cells, glial cells, or unusual hypodermal cells. Adding to our difficulties are the following possibilities. Some of these cells may have more important or obvious "neuronal" properties at certain stages in development, but those features may be absent in other stages. For example, a neuron may have a key role in axon guidance during embryonic development or in maintaining positions in larval development (Aurelio *et al.*, 2002; Hedgecock *et al.*, 1987; Wadsworth and Hedgecock, 1992; Wadsworth *et al.*, 1996). An important neuron in the male circuitry may also be present in the hermaphrodite, but lack any gender-specific synaptic relationships in that animal. Since we only have an incomplete knowledge of the male circuitry, such doubts are difficult to resolve. The cell's presence in the adult hermaphrodite can then be enigmatic.

The PVR neuron has shown inconsistent features in serial sections, so its normal cell fate is uncertain. This unpaired tail cell lies in the right lumbar ganglion and in some reconstructed adults has shown a posterior ciliated dendrite extending into the tail tip (Hall, 1977; Hall and Russell, 1991). In other animals the cell lacks the posterior cilium (White *et al.*, 1986). The neuron always extends an anterior axon to the nerve ring, and makes synapses there, but has few synaptic relationships in the tail circuits or along the ventral cord. Hence, it may be a sensory neuron or an interneuron.

Two "tail spike cells" are born in the tail tip and die later in embryogenesis (Sulston *et al.*, 1983; also see http://www.wormatlas.org/handbook/hypodermis/hypsupportother.htm). In one embryo at the two-fold stage sectioned by Sulston *et al.*, these cells appear to be fused into a syncytium and together they extend one ciliated process into the developing tail tip. The cells apparently have no axon, but do form electrical synapses to the neighboring hyp10 cell. It is not clear if this "cilium" has any signal transduction function, or if it is strictly a scaffold over which hyp10 develops the extended tail spike tissue that persists in larval and adult life.

The two XXX cells are born in the lips of the embryo and at first appear to adopt a hypodermal fate, apparently acting in concert with two hyp3 cells (Sulston *et al.*, 1983). After the body plan has been established, and at a time when other hypodermal cells are fusing into regional syncytia, the XXX cells fail to fuse, but begin crawling posterior-ward along with many neurons and glial cells of the anterior sensory anatomy. In the adult stage the XXX cells lie near the nerve ring, closely surrounded by neurons and glial cell bodies (White *et al.*, 1986). Their nuclei remain small and compact, and they extend several thin vesicle-containing processes towards the pseudocoelom, but form no obvious synapses. Ablation of the XXX cells in the first larval stage can cause animals to form dauer larvae (Ohkura *et al.*, 2003). Consistent with a role in regulating dauer entry, the XXX cells also express two genes of the dauer pathway, *daf-9* and *sdf-9* (Gerisch *et al.*, 2001; Jia *et al.*, 2002). The XXX cells also express the *npc-1* gene, a homologue of the Niemann-Pick type C disease gene predicted to function in cholesterol trafficking (Sym *et al.*, 2000). It appears likely that the XXX cells are acting in steroid metabolism and signaling. They may be equivalent to hormone-releasing cells in higher animals, perhaps more like gland cells than true neurons.

Six GLR cells lie just outside the nerve ring and extend thin scaffolding processes underlying the ring that may influence its development and final architecture. The GLR cells should not be listed as neurons. They were suggested to perform a glial function by White *et al.* (1986), based upon their close interactions with RME neurons, some sensory dendrites, and processes of the amphid sheath cells. Their processes do not form chemical synapses. The GLR processes do form gap junctions among themselves and with RME neurons. Over most of their length, GLR processes are more closely associated with muscle arms rather than with neuron processes. We classify them as part of the motor system, since they offer entry points and scaffolding processes for the arms of the head muscles to elaborate beneath the nerve ring (Altun *et al.*, 2005; also see http://www.wormatlas.org/handbook/mesodermal.htm/glr.htm).

The PDB neuron lies in the tail and extends a unique posterior-ward process far into the tail spike that travels past the phasmids before reversing to enter the dorsal nerve cord (Sulston *et al.*, 1980; White *et al.*, 1986). This nerve process is highly unusual and has no clear function. Perhaps the posterior process is stretch-sensitive, but there are no clear anatomical signs to support this. The cell forms two gap junctions to one motor neuron, but virtually no other synapses except the very few junctions to dorsal bodywall muscles that caused PDB to be labeled as a motor neuron (White *et al.*, 1986).

The AVG neuron is a unipolar neuron that sends its axon down the length of the ventral nerve cord, where it forms about 20 synapses. White *et al.* (1986) classified AVG as an interneuron. Oddly, some of these synaptic swellings seem to point away from neighboring neurites, contacting the basal lamina with no apparent post-synaptic partners. Wadsworth *et al.* (1996) and Durbin (1987) have established that AVG plays a key role as a pioneer axon for guiding the outgrowth of the embryonic ventral nerve cord. The AVG's synapses towards the pseudo-coelom suggest that the cell may also perform some neurohumoral role. Other neurohumoral cells could possibly include the NSM neurons of the pharynx (Albertson and Thomson, 1976) and the XXX cells (see the previous text). These cells also have vesicle release sites aimed towards the basal lamina and pseudo-coelom. Cells containing larger dense core vesicles might also be assigned to this class, regardless of their output zone. Finally, AVG also sends a process far posterior into the tail tip, which has been suggested to be a proprioceptor (White *et al.*, 1986).

The single PVT neuron of the preanal ganglion in the tail forms very few synapses. However, this neuron extends a process forward into the ventral nerve cord and secretes factors such as *zig* proteins that are necessary for keeping other ventral cord processes in their proper fascicles (Aurelio *et al.*, 2002). Loss of the PVT neuron by laser ablation causes similar axon placement defects as the *zig* mutants, and the key time of gene action is during larval stages, after initial nerve outgrowth is completed.

Four more neurons lying in the lumbar ganglia in the tail (ALNL/R, PLNL/R) offer a similar puzzle. They each extend anterior processes along the lateral bodywall in unique positions and, when they reach the nerve ring, they form few synapses. It is unclear if these neurons play an important role in any circuitry. It remains to be tested whether their role is to pioneer a dorso-lateral pathway for other processes such as the ALM dendrites (ALN) or for ventro-lateral pathways for the PLM dendrites and the ventral sublateral cords (PLN). All four cells also extend long posterior processes into the tail spike without any discernable function.

The PVN neurons extend processes all the way from their cell bodies in the lumbar ganglia to the nerve ring, running via the ventral cord. These cells are rather branchy in several locales, especially at the vulva and in the ring itself, but form few synapses. This again suggests a potential role in axon guidance. The PVNs might lead the way to the vulval muscles from the ventral cord, and perhaps help in some aspect of nerve ring formation, where they extend multiple branches within a confined track along the inner surface of the neuropil (White *et al.*, 1986).

Two CAN neurons lie laterally away from other neurons or ganglia. They each extend long neurites for the length of the midbody, from the nerve ring to the tail, in close association with the excretory canals and a few lateral neurites (ALA, PVD). These cells form almost no synapses anywhere along their length. However, when ablated with a laser, loss of the CAN neurons proved fatal to the animal (J. Sulston, 1980, unpublished results; White *et al.*, 1986). Thus, these cells provide some essential function and look much like other neurons, yet seem unimportant for synaptic wiring.

In the adult male tail, the glial cell PHso1 loses its socket cell (glial) function in the adult, and is replaced for that function by PHso2 (Sulston *et al.*, 1980). Curiously, the PHso1 cell now shows a variable number of basal bodies in the adult male, which would usually be a feature expected in a sensory neuron, but not a glial cell. Conversely, the PHso2 cell in the hermaphrodite is virtually extraneous, having only an incomplete glial wrapping of the phasmid sheath cell. Since the PHso1 cell shows no other signs of a neuronal cell fate (no axon, no synapses), it is still assigned to the glial class.

IV. Neuron Ultrastructure

Certain questions about the relevance of *C. elegans* neurons to other systems arise because of their small size and extreme simplicity in structure. We will briefly summarize several of these issues here, but other chapters in this volume will explore some of them in more detail.

A. Axons and Dendrites

1. Well-defined dendrites occur in which the process is specialized only for sensation, but has no synaptic interactions (Fig. 1). Sensory cells often form a separate axon that is involved in all the cell's synaptic interactions. When present, this axon process can stem from a local branch off the dendrite, rather than from an independent primary process from the soma.

2. Very few neurons have dendritic processes specialized for receiving synaptic inputs, with a separate axon process specialized for synaptic outputs. Examples where such segregation is maintained would include the DD and VD motor neurons and the RIP interneurons.

3. More often, synaptic zones for inputs and outputs are mixed along a single nerve process, which we might label a "neurite" rather than an axon or dendrite. Occasionally synaptic inputs and outputs occur on the same neuronal process as the ciliated ending (e.g., male tail).

4. "Axonal transport" occurs via microtubule-based motors as in higher animals (Hall and Hedgecock, 1991), but single motors seem to act bidirectionally at a uniform rate both in axons and in dendrites (Zhou *et al.*, 2001). This would seem to imply that the cytoskeletal organization of nematode neurites is more similar to vertebrate dendrites (where the microtubule bundles show mixed orientations) and less like vertebrate axons (where all microtubules are oriented with the "plus ends" away from the soma). Again this suggests that neuron processes are less specialized here than in higher animals.

5. There is no evidence for an axon hillock. Many neurons in higher animals have a specialized region at the base of the axon where the cell body summates the excitatory and inhibitory inputs to its dendrites, and initiates an action potential at the base of the axon (the hillock) which then passes down the axon to its synapses. There are specialized clusters of ion channels in the region of the hillock that are active in initiating the action potential, and their organization can be detected by freeze fracture (Matsumoto and Rosenbluth, 1985). Observations of *C. elegans* neurons by this technique have revealed no hint of such organization (Hall, 1987, unpublished observations).

6. In *C. elegans,* there is little evidence for action potentials in neurons, although pharyngeal muscles clearly exhibit them (Franks *et al.*, 2002; Raizen and Avery, 1994). In most neurons studied to date, it seems likely that graded potentials pass along neurites passively (Goodman *et al.*, 1998; See also Davis and Stretton, 1986). Since most neuronal processes have both synaptic inputs and outputs mixed without geographic segregation, local excitatory potentials (EPSPs) may develop and be transmitted from any point along the process. In sensory neurons with a distinct dendrite, there is a better chance that some region in the dendrite or near the soma may act as the generator of impulses in reaction to environmental stimuli. However, no structural evidence for such a specialized zone has been identified in *C. elegans* neurons.

7. There is no myelination along the axons. Neurons in higher species can often utilize glial wrappings (myelin) and clustering of intramembrane channels at myelin breaks (nodes) to increase the speed of conduction of signals (action potentials) along the length of individual axons (Peters *et al.*, 1990). In the nematode, axons have no glial wrappings, no nodes, and no apparent clustering of ion channels along the lengths of axons. And since they also appear to lack action potentials *per se*, there is little expectation that any structural feature could increase the conduction velocity of graded signals. Instead, the primary means to increase the gain on conducting these signals (or to minimize loss of signal) is that

the neurons maintain very high input resistance, probably by minimizing the channel density along their length (Goodman *et al.*, 1998).

B. The Neuron Growth Cone

During early outgrowth of neuron axons, pathfinding decisions are determined by a specialized "growth cone" at the leading edge of each neuronal process. The growth cone extends thin (actin-based) filopodia towards attractive cues, and withdraws or collapses filopodia when encountering repellent cues to cause reorientation of axon outgrowth. The anatomy of the growth cone has not been well characterized yet in nematode neurons *in vivo* or *in vitro*. However, it is clear that the rules for pathfinding are determined by the expression of specific receptors on the surface of the growth cone that respond to extracellular signals that lie on substrates such as the basal lamina or cell surfaces (Hedgecock *et al.*, 1987; Wadsworth *et al.*, 1996). A green fluorescent protein (GFP) marker has been used to label a growth cone and its filopodia in a larval animal, a stage when the growth cone is perhaps largest and most easily viewed in whole mount (Knobel *et al.*, 1999; Walthall and Chalfie, 1988). When followed in serial thin sections in the late embryo, growth cones are simple and quite small (Durbin, 1987; Norris *et al.*, 2000, unpublished observations).

C. Chemical Synapses

Although some chemical synapses in *C. elegans* involve simple "monadic" relationships where one presynaptic neuron contacts one postsynaptic neuron, monadic contacts are not the norm. Instead, a majority of synapses seem to be "polyadic"; they involve multiple postsynaptic partners across from a presynaptic active zone (Fig. 2). These have been called "dyadic synapses" where there are two postsynaptic partners, or "triadic synapses" where there are three. How many of the apparent postsynaptic partners are actually generating a postsynaptic potential (PSP) upon vesicle release? Since current methods for anatomy rarely show postsynaptic specialization, wiring diagrams are based upon apparent proximity of potential postsynaptic partners. Because of the remarkably small caliber of axonal processes in *C. elegans*, there are usually several more processes lying at short distances from the active zone (within 100 to 500 nm). We (and White *et al.*, 1976, 1978, 1983, 1986, 1992) only list partners where they appear to lie directly across the synaptic cleft from some portion of the presynaptic density. There is little structural information as yet regarding where vesicular quanta are actually released from the nematode synapse. As more becomes known in this regard, a reassessment of the published wiring diagrams may yield minor revisions. Neurophysiological measurements at wild type NMJs suggest that multiple postsynaptic partners do in fact generate EPSPs upon release of a single quanta at a chemical synapse (Liu *et al.*, 2005, unpublished observations).

D. NMJs

Virtually all contacts from neurons onto muscles occur at chemical NMJs. They look just like the polyadic synapses between neurons (Fig. 2). There are many muscle-muscle contacts via electrical synapses (gap junctions), particularly between muscle arms, but no obvious electrical contacts between neurons and muscles. Each NMJ involves an active zone in the plasma membrane of the presynaptic neuron process, where many synaptic vesicles are clustered close to the release site. The physical appearance of the presynaptic ending is no different from that at other chemical (nerve/nerve) synapses. On the postsynaptic side of the NMJ, one or more muscle arms extend close to the active zone, and another neuron process sometimes receives chemical input at the same site. It is difficult to visualize any postsynaptic specialization on the muscle arms, even by TEM. For some specialized muscles in the male tail, the presynaptic process may contact the muscle belly directly (PCB to anterior oblique muscle) or at the basal surface of the sarcomere (RnA to posterior outer longitudinal muscle; SPC to spicule protractor muscle), sometimes invaginating into the muscle cell (Lints *et al.*, 2005, unpublished observations; Sulston *et al.*, 1980). Neuronal contacts to pharyngeal muscles can also involve deep invaginations into the muscle cell, often occurring within the sarcomere region (Albertson and Thomson, 1976). Pharyngeal muscles do not form muscle arms, nor do some of the gender-specific muscles in the male tail.

E. Electrical Synapses

Electrical signaling can occur at gap junctions between two neurons. Gap junctions in the nematode involve homomeric or heteromeric intercellular junctions composed of innexin molecules (unrelated in sequence to vertebrate connexin proteins) (Phelan, 2000; Phelan and Starich, 2001). Each gap junction involves the close approach of the plasma membranes (extracellular gap is about 1 nm) to form an intercellular channel. By TEM, the gap junction is seen as a more electron dense region where the apposed membranes are straighter and evenly separated. Using the freeze fracture technique, the gap junction appears as a cluster of intramembrane particles and shallow pits, evenly spaced in an array, where the particles appear to be about 12 nm in diameter each (Hall, 1987). Whereas all vertebrate gap junctions involve "P-face" particles and "E-face pits," the nematode junctions vary according to tissue type. Neuronal gap junctions in the worm have roughly 50% particles and 50% pits on each fracture face. Neuronal gap junctions are generally very small in size, involving few channels (few particles and pits). In a thin section, they extend over such short regions of the apposed membranes that they are tricky to identify reliably. Improvements in fixation such as high-pressure freezing are making them somewhat easier to identify.

F. What Features Help to Identify a Sensory Ending?

Anatomical studies of sensory endings, when taken alone, have not been sufficient to clearly identify the modalities of all sensory neurons. Fortunately, the combination of genetic and laser ablation experiments, and the development of several behavioral paradigms have helped to substantiate the variety of the animal's sensory abilities, and to assign individual capabilities to certain sets of sense cells (Bargmann and Kaplan, 1998; Collet *et al.*, 1998; Ernstrom and Chalfie, 2002; Melkman and Sengupta, 2004). Many of these behavioral analyses and mutant studies are described elsewhere in this volume (see Chapters 2, 4, 5, and 6). Here, we will concentrate on the simple anatomical features that are known for the various sensory neurons. Table II gives an overview of the neurons in both genders that share these features. Figure 3 shows examples of some typical sensory specializations seen on nematode sensory dendrites.

Chemosensation is accomplished by a variety of neurons in the head and tail whose ciliated dendrites are exposed to the exterior environment through small holes in the body cuticle or fan cuticle (Table II). The number of chemicals that *C. elegans* can sense in its surroundings is larger than this cell list, and each cell probably expresses different receptors for several specific ligands (Bargmann and Horvitz, 1991; Bargmann and Mori, 1997; Hilliard *et al.*, 2002, 2004; Melkman and Sengupta, 2004). Conversely, the rather small group of sensors suggests that the animal's repertoire of different behavioral responses will be limited. There is some uncertainty that all neurons with exposed cilia are primary chemosensors or osmosensors. It is quite possible that some of these neurons are more sensitive to mechanical deformation, as demonstrated already for ASH (Kaplan and Horvitz, 1993). Since chemoreceptors are deployed in both the head and tail sensilla, some signals can be compared at the two ends of the animal when the animal decides upon a response (Hilliard *et al.*, 2002).

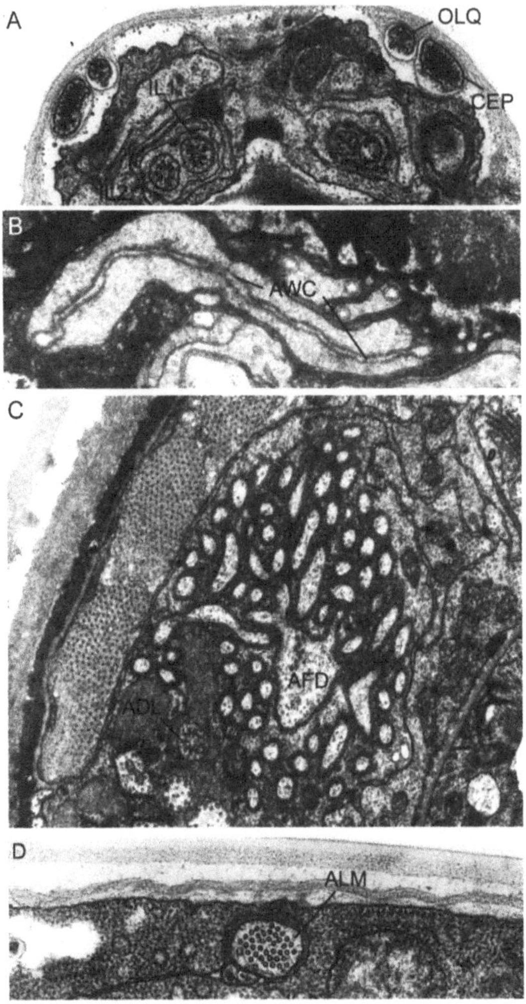

FIG. 3. Sensory dendrite specializations. (A) Two types of mechanoreceptors are shown embedded in the dorsal lip cuticle, OLQ and CEP (right members of a bilateral pair are marked). Two inner labial cilia, IL1 and IL2, are marked on the left side. The tip of the IL2 cilium opens outside of the cuticle through a narrow channel (not shown) and this cell is chemosensory. (B) The tip of the AWC cilium broadens into a very wide fan-like structure, embedded into the lip tissue. This neuron is sensitive to volatile odorants that can penetrate the cuticle and surrounding tissue. (C) The base of the left AFD cilium is marked, and is surrounded by dozens of fine villi that branch from the cilium. The villi are embedded in thin wrappings of the amphid sheath cell. This cell is thermosensory. The cilium of the left ADL neuron is marked within the amphid channel, surrounded by extracellular "ground substance." This chemosensory dendrite is exposed to the outside through a large hole in the lip cuticle (not shown). (D) The dendrite of the right ALM neuron is shown at a cross-section through the anterior midbody region. This touch-sensitive dendrite is filled with distinctively

Odorant receptors have ciliated dendrites that terminate without direct exposure to the outside. Three known odorant receptors—AWA, AWB, and AWC—have broad-flattened processes embedded into the amphid sheath in the nose (Sagasti *et al.*, 1999). Oxygen tension (and perhaps NO, SO, and so on) is sensed by ciliated receptors lying deep inside the body cavity and one in the tail (Gray *et al.*, 2004). Again the repertoire of responses to different odorants is going to be limited by the worm's simple circuitry. Multiple signals received by the AWC dendrite are all thought to promote attraction towards the signals, whereas all signals received by the AWB dendrite are expected to promote repulsion (Bargmann, 1993; Melkman and Sengupta, 2004; Troemel *et al.*, 1997).

Temperature sensation is accomplished by buried AFD nerve endings in the nose. Mutations that interfere with development of this sense ending also abolish normal thermotaxis (Hedgecock and Russell, 1975). As previously noted, these ciliated endings have highly branched villi beyond the basal body, embedded in the sheath cell of the amphid. A variety of experimental evidence predicts that a second thermosensor is present, but its identity has yet to be proven (Mori, 1999; Mori and Ohshima, 1995).

Sensation of mechanical stimuli is of key importance to the animal's survival, and several types of receptors are distributed along the body (see Chapter 6). "Light touch" onto the body (such as by a fine hair) is well detected all over the body surface by specialized dendrites containing a distinctive bundle of large diameter microtubules (Chalfie and Thomson, 1982). This "touch dendrite" is firmly linked to the hypodermis, its basal lamina, and to the cuticle (Emtage *et al.*, 2004) so that local deformation of the cuticle is quickly detected regionally by separate dendrites covering the anterior (ALM, AVM) or posterior (PLM, PVM) body. Harsh touch (such as by a sharp wire) is also detected over the whole body, perhaps by unspecialized thin processes of the PVD cells (Treinin, 2005, personal communication; Tobin and Bargmann, 2004). Light touch to the nose is sensed primarily by the ASH neurons of the amphids (Kaplan and Horvitz, 1993; Tobin and Bargmann, 2004). Two pairs of mechanosensors form small cuticle nubbins along the alae, one pair just behind the head (ADE) and another pair near the tail (PDE) (Perkins *et al.*, 1986). One can imagine that these cells are in good position to help the animal detect which side of its body is lying against the substrate, since it typically lies on one side or the other. Similar cuticle-embedded receptors are concentrated at the head and tail, giving the animal several means to detect

---

large microtubules, and is linked to the overlying hypodermis via extracellular mantle protein, and indirectly linked to the cuticle by various *mec* proteins. Images for panels A, B, and C are from the MRC/LMB archives (courtesy of John White and Jonathan Hodgkin). Panel D is from the Hall lab archives. All represent wild-type adults. Magnifications are ~30,000X for panels A to C and 60,000X for panel D.

obstacles during forward or backward motion. The adult male has even more mechanoreceptors in the tail that are used in detecting the presence of the hermaphrodite during mating, including nine pairs of sensors in the rays.

Except for the PVD neurons previously mentioned, all of the apparent sensory neurons listed have ciliated dendrites or specialized touch dendrites that are easily distinguished just by their fine structure. Many other "bare-wire" processes have been noted in thin section reconstructions that are suggestive of a role in sensation due to their special positions, and often due to their extremely long lengths without any other apparent purpose (non-synaptic). Several classes of neurons send unspecialized processes far into the tail tip, a structure that undergoes severe deformation whenever the animal moves backwards (Hall, 1977; Hall and Russell, 1991). Since there is only one ciliated dendrite present this far posterior (PVR), and that cilium is not always present, the processes of PHC, PDB, and AVG are regarded as likely proprioceptors (detecting internal body deformation). The SPC processes running deep into spicule muscles are again presumed to be sensory (discussed previously). Several other candidate bare-wire receptors are noted in Table II, most thought to detect local stretching of the bodywall during the animal's motion. Another large set of potential bare wire sensors are in the pharynx, each glued to its cuticle by prominent specialized junctions; these may be in position to detect food within the pharynx (Albertson and Thomson, 1976).

V. Methods for Analytical Study of Neurons in *C. elegans*

A. EM Methods

1. *TEM*

Sydney Brenner had this technique in mind when he first selected *C. elegans* as an animal model (Brenner, 1974). He hoped that the precision and high resolution of TEM would permit the complete analysis of neuronal circuits among identified cells, and the analysis of mutant strains in which those circuits were miswired. When combined with serial thin section reconstruction, TEM has substantially succeeded in the first of Brenner's aims, but reconstruction methods are still so cumbersome that few mutants are understood in terms of their wiring patterns. Variations in fixation and embedding techniques for the nematode have slowly evolved to ease the analysis of smaller larvae, embryos, and early mutant phenotypes (Hall, 1995; Paupard *et al.*, 2001). Current protocols are available on the Anatomical Methods section at www.wormatlas.org. Methods for immunolabeling neurons in thin sections have been improved recently (see the following text) to permit the localization of individual proteins in the vicinity of neuron synapses. High-pressure freeze (HFP) techniques (Kosinski *et al.*, 2005; McDonald, 1994; Rolls *et al.*, 2002; Rostaing *et al.*, 2004) improve the power of TEM to distinguish subtle changes in synaptic structures in behavioral mutants.

Much of the early TEM study of *C. elegans* was geared more towards the efficient identification of neuron shapes and arbors, using osmium-only fixations that highlighted membrane borders but failed to adequately preserve cytoplasmic contents (Albertson and Thomson, 1976; Ward *et al.*, 1975; Ware *et al.*, 1975; White *et al.*, 1976; reviewed by Hall, 1995). Two-step fixations where an aldehyde fixation precedes the osmium stain are an improvement for cytoplasmic features, but the contents of synaptic swellings may still degrade for many minutes before the first fixative enters the animal. The HPF technique was first used on *C. elegans* by Colleen Lavin in the White lab, primarily with the goal of exploring events in early embryos (J. White, 1999, personal communication). HPF followed by a freeze substitution fixation is believed to fix the cell membranes, microtubules, and synaptic vesicle contents within a few microseconds and, therefore, can preserve a much more representative view of the dynamics of synaptic organization and vesicle release (Rolls *et al.*, 2002; Rostaing *et al.*, 2004). Because better HPF protocols for neuron preservation are still in development for *C. elegans*, we expect that much new TEM information on individual synapses and mutant phenotypes remains to be uncovered in the next few years.

2. *Other EM Technologies*

Scanning electron microscopy (SEM) has been less helpful in studying neurons in *C. elegans*. The technique helps to view the outside of the animal, so it is best suited for viewing details of cuticle organization. This has been of some interest in viewing the location and size of some cuticle-associated dendritic endings (amphids, deirids, phasmids, and male rays) (Riddle, 1988).

Freeze fracture and freeze etch techniques (Hall, 1987, 1995; Hall *et al.*, 1999) have been used rather little in *C. elegans*, but can provide high-resolution information about the distribution of membrane channels in neuronal membranes, such as gap junctions and perhaps those associated with the synaptic active zone. These methods are not especially suitable for recognizing which exact neuron membrane is being viewed, and this limits its power in mutant analyses.

Electron tomography is now capable of delivering high-resolution three-dimensional images from thick sections of fast-frozen tissue in *C. elegans* (Muller-Reichert *et al.*, 2003). By using a high-resolution tilting stage on the electron microscope, tomographic reconstruction of a whole neuron process or synapse can be achieved within that section, allowing the user to see relationships of whole organelles to their environment (Marsh *et al.*, 2001; McEwen and Marko,

2001; McIntosh *et al.*, 2005). This technique is becoming more widely available due to recent improvements in tissue preparation, microscope equipment, and computerized aids to reconstruction.

Energy-filtered TEM is also feasible to demonstrate the exact disposition of certain atoms (phosphorus imaging, calcium imaging) at high resolution in *C. elegans* using a modern electron microscope having an "energy filter" built into the column (Leapman *et al.*, 2004). Because these microscopes are still rather uncommon, little has been achieved yet in the nematode. However, the animal's small size and excellent definition of all cell types in thin sections makes this technique very attractive.

3. *Immunocytochemistry at the TEM Level*

Improvements on the immunoEM technique made over the past few years are yielding better resolution of neuronal proteins. Microwave fixation and embedment in acrylic resins such as LR White or LR Gold helps to prepare large numbers of specimens for post-embedding immunolabeling (Paupard *et al.*, 2001). Approximately one-half of the available antibodies seem to have some chance for success when applied to plastic thin sections. The other one-half may fail due to degradation of the epitope during fixation or dehydration of the tissue. However, microwave fixation does not offer good resolution of neurons or synapses. HPF fixation followed by freeze substitution into either LR Gold or Lowicryl resins has greatly improved neuronal preservation for nematodes and other invertebrates (Howe *et al.*, 2001; Kosinski *et al.*, 2005; McDonald, 1994; Rappleye *et al.*, 1999). New protocols have been tested by Bessereau and colleagues that have produced high-resolution images of several synaptic proteins (Rostaing *et al.*, 2004; Yeh *et al.*, 2005). Again the chances for success are roughly 50% or better, depending upon the epitope, which suggests that epitope degradation remains an obstacle. It is also possible to collect frozen thin sections of the animal and conduct immunogold cytochemistry on these sections without resort to plastic embedment (Sato *et al.*, 2005; Selkirk *et al.*, 1991). The frozen section method has seemed more difficult to master, and has been used much less often on nematodes.

4. *Serial Thin Section Reconstruction*

Because the early era of reconstruction efforts occurred before the availability of inexpensive computing power, most work was done either using film techniques (Ware and LoPresti, 1975; Ware *et al.*, 1975) or using stacks of prints and colored marking pens to annotate the prints. The procedure was slow and arduous, so that the final accomplishments of such works as "Mind of the Worm" by White and colleagues (1986) are truly monumental. Very similar efforts were carried out at MRC (Cambridge, England) and Caltech (Pasadena, CA) over a period of some 15 years, involving about 15 scientists and technicians to fully trace the nervous anatomy of the adult hermaphrodite and the tail of the adult male. Both groups produced essentially the same results over limited regions (compare Ward *et al.*, 1975, and Ware *et al.*, 1975), with each project focusing on the neurons and processes of just a very few animals in total. The MRC group succeeded much further in enumerating the wiring diagram for the animal, based upon overlapping data from five animals (White *et al.*, 1986). The wiring diagram for the male tail was still incomplete when published (Sulston *et al.*, 1980), principally based upon reconstruction of one exceptionally well-fixed specimen. Reconstructions of limited portions of the *unc-4* nerve cords from three animals showed a singular change in the wiring pattern of certain interneurons and motor neurons compared to the wild-type pattern (White *et al.*, 1992). The wiring diagrams of several more *unc* mutations have been analyzed using the same hand methods on prints by the MRC group, but remain unpublished.

B. Light Microscopy Methods

1. *Nomarski/Hofman Optics/DIC*

Variations on differential interference microscopy (DIC) are one of the most frequently used techniques to visualize cells and cell nuclei in live animals (Sulston and Hodgkin, 1988). Because *C. elegans* is virtually clear and small enough to see through, these techniques add enough contrast for one to focus separately on very narrow depths of field and to improve the details of cytoplasmic contents (nucleus, nucleolus) of individual neurons. Using these methods, one can follow the patterns of cell divisions as the animal matures, and produce a lineage map of any portion of the anatomy at any age (Sulston and Horvitz, 1977). Cells undergoing morphogenesis can be distinguished from cells undergoing programmed cell death, and subtle changes in cell position can be tracked as tissue development continues. Mutations that cause neuronal swelling (necrosis) can be imaged in live animals without need for a marker or stain (Driscoll and Chalfie, 1991; Yochem *et al.*, 2005). These techniques by themselves are less practical for exploring the nature of neuronal processes, since the individual axons and dendrites are too narrow to be distinguished without some type of vital staining. But when combined with intracellular markers such as GFP or DiO (see the following text), one can monitor the time course of neuron development and identify mutant phenotypes at low resolution. Alternately, in a *clr-1* background, one can obtain vivid live images of identified neurons and their processes (Hedgecock *et al.*, 1990).

2. *Confocal Microscopy/Two Photon Confocal Microscopy*

Improvements in confocal microscopy technology, often spurred by its utility for *C. elegans* studies (Eliceiri *et al.*, 2003; Mohler and White, 1998a,b; Thomas and White, 1998), have allowed high-gain, high-magnification light microscopy to help in visualizing single cells and labeled cell parts, often in conjunction with fluorescent gene expression markers such as GFP. Although originally developed for use in fixed specimens, where the confocal technology permitted exceptional views of tissues within extremely narrow depths of focus, more recent adaptations have also permitted high-resolution views of live specimens. Combined with caged stains that can be released by laser light, or with vital dyes that shift their emitted wavelength depending upon physiological conditions, "real time" confocal methods permit one to view changes in tissue physiology or even the transport of intracellular particles (along cilia or axons) in the millisecond timeframe (Hilliard *et al.*, 2005; Kerr *et al.*, 2000; Schafer *et al.*, 2003; Suzuki *et al.*, 2003; Zhou *et al.*, 2001).

"Deconvolution" software has been used to render detailed three-dimensional neuron shapes from confocal data stacks (Coburn and Bargmann, 1996; Thomas and White, 1998). This approach uses statistical methods to identify and reject out-of-focus image haze to refine the cell profile in thin focal planes, so that more accurate representations can be achieved in a focus-through image stack from a whole mount worm. This avoids the use of TEM and serial section methods, which are much slower and perhaps more expensive. More recently, a multi-color cell labeling scheme has been developed to view the three-dimensional relations of five parallel axons in the ventral cord from confocal images, again to avoid using TEM and serial sections (Hutter, 2004).

Methods for gluing the live animal to a microscope slide, combined with intracellular calcium-sensitive fluorophores have permitted the real-time analysis of neuron activity at the cell soma (Hilliard, 2005; Kerr *et al.*, 2000; Suzuki *et al.*, 2003). The animal can be bathed in soluble repellants or attractants, or stroked gently to activate sensory responses. When combined with low-power imaging of the whole animal, it is also possible to match the animal's ongoing behavior to the intracellular activity of a known neuron (S. Lockery, 2005, personal communication).

3. *DiI/DiO and FITC Markers*

These vital dyes will cross the plasma membrane of ciliated dendrites to fill the cell body and all processes of sensory neurons if their cilium is exposed to the exterior environment of the nematode (Lundquist and Herman, 1994; Perkins *et al.*, 1986). The live worms can swim in this staining solution, and after rinsing to remove excess stain from the animal's surface, one can visualize the number of neurons that form functional dendrites in the amphids and phasmids, and the route by which they send axons towards the nerve ring or elsewhere. This has proven useful in assessing mutant phenotypes, and as the basis for genetic screens (Collett *et al.*, 1998; Hedgecock *et al.*, 1990). The method does not help for most neurons, which lack the needed ciliated dendrite to be filled, and is too low in resolution to be very useful in detecting synaptic swellings along individual processes, particularly because it tends to indiscriminately label many similar sensory axons in parallel, whose synaptic arbors are closely bundled together.

4. *GFP/YFP/CFP Markers*

Transgenic markers that specifically label neurons with GFP or similar colored protein markers (YFP, CFP, dsRED) in live nematodes have revolutionized our ability to assess the birth, morphogenesis and decline of specific neurons during the life of the animal (Chalfie *et al.*, 1994; Miller *et al.*, 1999). By marking different neurons with several different colors in the same individual, one can view physical relations between identified neuron processes (Hutter, 2004). Because these protein markers fluoresce even in the living animal, they are ideal for following developmental changes in cell shape. By adding the GFP moiety to a functional neuronal protein, the label can be transported to the normal site of that protein and label just a particular organelle or even single synapses (Zhen and Jin, 1999). This has proven useful in assessing mutant phenotypes, and as the basis for genetic screens for mutant suppressors and revertants (Crump *et al.*, 2001).

5. *Antibody Staining*

Whole mount immunolabeling of neurons or neuronal organelles has proven a powerful means to compare the development of the nervous system in all stages of development, and in comparing defects in genetic mutations to normal animals (Finney and Ruvkun, 1990). Miller and Shakes (1995) have reviewed the methods both for antibody preparation and for specimen preparation techniques. Since the eggshell and cuticle coverings offer rather different obstacles to efficient staining as the animal matures, these techniques vary when treating embryos, young larvae, or adults. Conversely, the same methods work well for neuronal and non-neuronal epitopes.

6. *Laser Ablation*

Single neurons can be selectively killed with a laser microbeam while under view on the light microscope stage (Bargmann and Avery, 1995; White and Horvitz, 1979). The high selectivity of the method has proven extremely powerful in analyzing the functions of individual neurons and neuron classes. It is also feasible to kill off multiple cells in a single animal, so that the role of the neurons in a simple circuit can be assessed accurately and reproducibly (Avery and Horvitz, 1989).

Laser ablations remain a rather blunt instrument in testing the wiring diagram, or in testing the role of individual neurons. In the absence of more direct measures of neuronal activity (excitation or inhibition) it has been mostly a thought exercise to predict the animal's behavior based upon the wiring diagram. Improved methods for cell monitoring in the live animal, such as monitoring calcium waves inside single identified neurons are rapidly improving this situation (see the previous text). Combined with mutant analysis, these methods will empower neuroscientists to make more testable predictions about how cell-cell signaling operates within the predicted wiring diagram to control worm behavior (Dunn *et al.*, 2004; Whittaker and Sternberg, 2004).

References

Albertson, D. G., and Thomson, N. (1976). The pharynx of *C. elegans*. *Phil. Trans. Roy. S. Lond.* **275,** 299–325.

Altun, Z. F., Lints, R., and Hall, D. H. (2005). Atlas of *C. elegans* anatomy. An illustrated handbook. www.wormatlas.org/handbook/contents.htm.

Aurelio, O., Hall, D. H., and Hobert, O. (2002). Immunoglobulin-domain proteins required for maintenance of ventral nerve cord organization. *Science* **295,** 686–690.

Avery, L., and Horvitz, H. R. (1989). Pharyngeal pumping continues after laser killing of the pharyngeal nervous system of *C. elegans*. *Neuron* **3,** 473–485.

Avery, L., and Thomas (1997). Feeding and defecation. *In* "*C. elegans*" (D. L. Riddle, T. Blumenthal, B. J. Meyer, andJ. R. Priess, Eds.), Vol. II, pp. 679–716. Cold Spring Harbor Laboratory Press, New York.

Bargmann, C. I. (1993). Genetic and cellular analysis of behavior in *C. elegans*. *Ann. Rev. Neurosci.* **16,** 47–71.

Bargmann, C. I. (1998). Neurobiology of the *Caenorhabditis elegans* genome. *Science* **282,** 2028–2033.

Bargmann, C. I., and Horvitz, H. R. (1991). Chemosensory neurons with overlapping functions direct chemotaxis to multiple chemicals in *C. elegans*. *Neuron* **7,** 729–742.

Bargmann, C. I., and Avery, L. (1995). Laser killing of cells in *Caenorhabditis elegans*. *In* "*C. elegans*: Modern Biological Analysis of an Organism" (H. F. Epstein and D. C. Shakes, Eds.), pp. 225–250. Academic Press, New York.

Bargmann, C. I., and Mori, I. (1997). Chemotaxis and thermotaxis. *In* "*C. elegans*" (W. Wood, Ed.), Vol. II, pp. 717–737. Cold Spring Harbor Laboratory Press, New York.

Bargmann, C. I., and Kaplan, J. M. (1998). Signal transduction in the nervous system. *Ann. Rev. Neurosci.* **21,** 279–308.

Brenner, S. (1974). The genetics of *Caenorhabditis elegans*. *Genetics* **77,** 71–94.

Bullock, T., and Horridge, G. (1965). "Structure and Function in the Invertebrate Nervous System." WH Freeman, San Francisco.

Burr, A. H. J. (1985). The photomovement of *C. elegans*, a nematode that lacks ocelli. Proof that the response to light is not radiant heating. *Photochem. Photobiol.* **41,** 577–582.

Cassata, G., Kagoshima, H., Andachi, Y., Kohara, Y., Durrenburger, M. B., Hall, D. H., and Burglin, T. R. (2000). The Lim homeobox gene *ceh-14* confers thermosensory function to the AFD neurons in *Caenorhabditis elegans*. *Neuron* **25,** 587–597.

Chalfie, M., and Thomson, J. N. (1982). Structure and functional diversity in the neuronal microtubules of *C. elegans*. *J. Cell Biol.* **93,** 15–23.

Chalfie, M., Sulston, J. E., White, J. G., Southgate, E., Thomson, J. N., and Brenner, S. (1985). The neural circuit for touch sensitivity in *Caenorhabditis elegans*. *J. Neurosci.* **5,** 956–964.

Chalfie, M., and White, J. (1988). The nervous system. *In* "The Nematode *Caenorhabditis elegans*" (W. B. Wood, Ed.), pp. 337–391. Cold Spring Harbor Laboratory Press, New York.

Chalfie, M., Tu, Y., Euskirchen, G., Ward, W., and Prasher, D. (1994). Green fluorescent protein as a marker for gene expression. *Science* **263,** 802–805.

Chao, M. Y., Komatsu, H., Fukuto, H. S., Dionne, H. M., and Hart, A. C. (2004). Feeding status and serotonin rapidly and reversibly modulate a *Caenorhabditis elegans* chemosensory circuit. *PNAS* **101,** 15512–155127.

Chen, B. L., Hall, D. H., and Chklovskii, D. B. (2005). Wiring optimization can relate neuronal structure and function. Submitted to PNAS.

Cheung, B. H. H., Arellano-Carbajal, F., Rybicki, I., and De Bono, M. (2004). Soluble guanylate cyclases act in neurons exposed to the body fluid to promote *C. elegans* aggregation behavior. *Curr. Biol.* **14,** 1105–1111.

Coburn, C. M., and Bargmann, C. I. (1996). A putative cyclic nucleotide-gated channel is required for sensory development and function in *C. elegans*. *Neuron* **17,** 695–706.

Collet, J., Spike, C. A., Lundquist, E. A., Shaw, J. E., and Herman, R. K. (1998). Analysis of *osm-6*, a gene that affects sensory cilium structure and sensory neuron function in *Caenorhabditis elegans*. *Genetics* **148,** 187–200.

Crump, J. G., Zhen, M., Jin, Y. S., and Bargmann, C. I. (2001). The SAD-1 kinase regulates presynaptic vesicle clustering and axon termination. *Neuron* **29,** 115–129.

Dal Santo, P., Logan, M. A., Chisholm, A. D., and Jorgensen, E. M. (1999). The inositol triphosphate receptor regulates a 50-second behavioral rhythm in *C. elegans*. *Cell* **98,** 757–767.

Davis, R. E., and Stretton, A. O. (1986). The motor nervous system of *Ascaris*: Electrophysiology and anatomy of the neurons and their control by neuromodulators. *Parasitology* **113**(Suppl), S97–S117.

Driscoll, M. A., and Chalfie, M. (1991). The *mec-4* gene is a member of a family of *Caenorhabditis elegans* genes that can mutate to induce neuronal degeneration. *Nature* **349,** 588–593.

Dunn, N. A., Lockery, S. R., Pierce-Shinomura, J. T., and Connery, J. S. (2004). A neural network model of chemotaxis predicts function of synaptic connections in the nematode *C. elegans*. *J. Computational Neurosci.* **17,** 137–147.

Durbin, R. M. (1987). "Studies on the development and organisation of the nervous system of *Caenorhabditis elegans*". Ph.D. dissertation. University of Cambridge, England.

Eliceiri, K. W., Fan, C. H., Lyons, G. E., and White, J. G. (2003). Analysis of histology specimens using lifetime multiphoton microscopy. *J. Biomed. Opt.* **8,** 376–380.

Emmons, S. W., and Sternberg, P. W. (1997). Male development and mating behavior. *In* "The Nematode *C. elegans*" (D. L. Riddle, T. Blumenthal, B. J. Meyer, and J. R. Priess, Eds.), Vol. II, pp. 295–334. Cold Spring Harbor Laboratory Press, New York.

Emtage, L., Gu, G., Hartwieg, E., and Chalfie, M. (2004). Extracellular proteins organize the mechanosensory channel complex in *C. elegans* touch receptor Neurons. *Neuron* **44,** 795–807.

Ernstrom, G. G., and Chalfie, M. (2002). Genetics of sensory mechanotransduction. *Ann. Rev. Genetics* **36,** 411–453.

Finney, M., and Ruvkun, G. B. (1990). The *unc-86* gene product couples cell lineage and cell identity in *C. elegans*. *Cell* **63,** 895–905.

Franks, C. J., Pemberton, D., Vinogradova, I., Cook, A., Walker, R. J., and Holden-Dye, L. (2002). Ionic basis of the resting potential and action potential in the pharyngeal muscle of *Caenorhabditis elegans*. *J. Neurophysiol.* **87,** 954–961.

Gerisch, R., Weitzel, C., Kober-Eisermann, C., Rottiere, V., and Antebi, A. (2001). A hormonal signaling pathway influencing *C. elegans* metabolism, reproductive development, and lifespan. *Developmental Cell* **1,** 841–851.

Gomez, M., de Castro, E., Guarin, E., Sasakura, H., Kuhara, A., Mori, I., Bartfai, T., Bargmann, C. I., and Nef, P. (2001). Ca2 signaling via the neuronal calcium sensor-1 regulates associative learning and memory in *C. elegans*. *Neuron* **30,** 241–248.

Goodman, M. B., Hall, D. H., Avery, L., and Lockery, S. R. (1998). Active currents regulate sensitivity and dynamic range in *C. elegans* neurons. *Neuron* **20,** 763–772.

Gray, J. M., Karow, D. S., Lu, H., Chang, A. J., Chang, J. S., Ellis, R. E., Marletta, M. A., and Bargmann, C. I. (2004). Oxygen sensation and social feeding mediated by a *C. elegans* guanylate cyclase homologue. *Nature* **430,** 317–322.

Halevi, S., McKay, J., Palfreyman, M., Yassin, L., Eshel, M., Jorgensen, E., and Treinin, M. (2002). The *C. elegans ric-3* gene is required for maturation of nicotinic acetylcholine receptors. *EMBO J.* **21,** 1012–1020.

Hall, D. H. (1977). "The posterior nervous system of the nematode *Caenorhabditis elegans.*" Ph.D. thesis. Caltech, Pasadena.

Hall, D. H. (1987). Freeze fracture and freeze etch studies of the nematode *Caenorhabditis elegans. N.Y. Acad. Sci.* **494,** 215–217.

Hall, D. H. (1995). Electron microscopy and 3D image reconstruction. *In* "*C. elegans*: Modern Biological Analysis of an Organism" (H. F. Epstein and D. C. Shakes, Eds.), pp. 395–436. Academic Press, New York.

Hall, D. H., and Russell, R. L. (1991). The posterior nervous system of the nematode *Caenorhabditis elegans*: Serial reconstruction of identified neurons and complete pattern of synaptic interactions. *J. Neurosci.* **11,** 1–22.

Hall, D. H., and Hedgecock, E. M. (1991). Kinesin-related gene unc-104 is required for axonal transport of synaptic vesicles in *C. elegans. Cell* **65,** 837–847.

Hall, D. H., Winfrey, V. P., Blauer, G., Hoffman, L., Furuta, T., Rose, K. L., Hobert, O., and Greenstein, D. (1999). Ultrastructural features of the adult hermaphrodite gonad of *C. elegans*: Relations between the germ line and soma. *Dev. Biol.* **212,** 101–123.

Hedgecock, E. M., and Russell, R. L. (1975). Normal and mutant thermotaxis in the nematode *C. elegans. PNAS* **72,** 4061–4065.

Hedgecock, E. M., Culotti, J. G., Hall, D. H., and Stern, B. D. (1987). Genetics of cell and axon migrations in *C. elegans. Development* **100,** 365–382.

Hedgecock, E. M., Culotti, J. G., and Hall, D. H. (1990). The *unc-5, unc-6,* and *unc-40* genes guide circumferential migrations of pioneer axons and mesodermal cells on the epidermis in *C. elegans. Neuron* **4,** 61–85.

Hilliard, M., Bargmann, C. I., and Bazzicalupo, P. (2002). *C. elegans* responds to chemical repellents by integrating sensory inputs from the head and the tail. *Curr. Biol.* **12,** 730–734.

Hilliard, M. A., Bergamasco, C., Arbucci, S., Plasterk, R. H. A., and Bazzicalupo, P. (2004). Worms taste bitter: ASH neurons, QUI-1, GPA-3 and ODR-3 mediate quinine avoidance in *Caenorhabditis elegans. EMBO J.* **23,** 1101–1111.

Hilliard, M. A., Apicella, A. J., Kerr, R., Suzuki, H., Bazzicalupo, P., and Schafer, W. R. (2005). *In vivo* imaging of *C elegans* ASH neurons: Cellular response and adaptation to chemical repellents. *EMBO J.* **24,** 63–72.

Howe, M., McDonald, K. L., Albertson, D. G., and Meyer, B. J. (2001). HIM-10 is required for kinetochore structure and function on *Caenorhabditis elegans* holocentric chromosomes. *J. Cell Biol.* **153,** 1227–1238.

Hutter, H. (2004). Five color *in vivo* imaging of neurons in *Caenorhabditis elegans. J. Microscopy* **215,** 213–218.

Jia, K., Albert, P. S., and Riddle, D. L. (2002). DAF-9, a cytochrome P450 regulating *C. elegans* larval development and adult longevity. *Development* **129,** 221–231.

Kaplan, J. M., and Horvitz, H. R. (1993). A dual mechanosensory and chemosensory neuron in *Caenorhabditis elegans. PNAS* **90,** 2227–2231.

Kerr, R. A., Lev-Ram, V., Baird, G., Vincent, P., Tsien, R. Y., and Schafer, W. R. (2000). Optical imaging of calcium transients in neurons and pharyngeal muscle of *C elegans. Neuron* **26,** 583–594.

Knobel, K. M., Jorgensen, E. M., and Bastiani, M. J. (1999). Growth cones stall and collapse during axon outgrowth in *C. elegans. Development* **126,** 4489–4498.

Kosinski, M., McDonald, K., Schwartz, J., Yamamoto, I., and Greenstein, D. (2005). *C. elegans* sperm bud vesicles to deliver a meiotic maturation signal to distant oocytes. *Development* **132,** 3357–3369.

Leapman, R. D., Kocsis, E., Zhang, G., Talbot, T. L., and Laquerriere, P. (2004). Three-dimensional distributions of elements in biological samples by energy-filtered tomography. *Ultramicroscopy* **100,** 115–125.

Liu, D. W. C., and Thomas, J. H. (1994). Regulation of a periodic motor program in *C. elegans. J. Neuroscience* **14,** 1953–1962.

Lundquist, E. A., and Herman, R. K. (1994). The *mec-8* gene of *C. elegans* affects muscle and sensory neuron function and interacts with three other genes: *unc-52*, *smu-1* and *smu-2*. *Genetics* **138,** 83–101.

Marsh, B. J., Mastronarde, D. N., Buttle, K. F., Howell, K. E., and McIntosh, J. R. (2001). Organellar relationships in the Golgi region of the pancreatic beta cell line, HIT-T15, visualized by high resolution electron tomography. *PNAS* **98,** 2399–2406.

Matsumoto, E., and Rosenbluth, J. (1985). Plasma membrane structure at the axon hillock, initial segment and cell body of the frog dorsal root ganglion cells. *J. Neurocytol.* **14,** 731–747.

McDonald, K. L. (1994). Electron microscopy and EM immunocytochemistry. *Methods Cell Biol.* **44,** 411–444.

McEwen, B. F., and Marko, M. (2001). The emergence of electron tomography as an important tool for investigating cellular ultrastructure. *J. Histochem. Cytochem.* **49,** 553–563.

McIntire, S. L., Jorgensen, E. M., and Horvitz, H. R. (1993). Genes required for GABA function in *Caenorhabditis elegans*. *Nature* **364,** 334–337.

McIntosh, R., Nicastro, D., and Mastronarde, D. (2005). New views of cells in 3D: An introduction to electron tomography. *Trends Cell Biol.* **15,** 43–51.

Melkman, T., and Sengupta, P. (2004). The worm's sense of smell. Development of functional diversity in the chemosensory system of *Caenorhabditis elegans*. *Dev. Biol.* **26,** 302–319.

Miller, D. M., and Shakes, D. C. (1995). Immunofluorescence microscopy. *In* "*C. elegans*: Modern Biological Analysis of an Organism" (H. F. Epstein and D. C. Shakes, Eds.), pp. 365–395. Academic Press, New York.

Miller, D. M., Desai, N. S., Hardin, D. C., Piston, D. W., Patterson, G. H., Fleenor, J., Xu, S. Q., and Fire, A. (1999). Two-color GFP expression system for *C. elegans*. *Biotechniques* **26,** 914–918.

Mohler, W. A., and White, J. G. (1998a). Stereo-4-D reconstruction and animation from living fluorescent specimens. *Biotechniques* **24,** 1006–1012.

Mohler, W. A., and White, J. G. (1998b). Multiphoton laser scanning microscopy for four-dimensional analysis of *Caenorhabditis elegans* embryonic development. *Optics Express* **3,** 325–331.

Mori, I., and Ohshima, Y. M. (1995). Neural regulation of thermotaxis in *Caenorhabditis elegans*. *Nature* **376,** 344–348.

Mori, I. (1999). Genetics of chemotaxis and thermotaxis in the nematode *Caenorhabditis elegans*. *Ann. Rev. Genetics* **33,** 399–422.

Muller-Reichert, T., Hohenberg, H., O'Toole, E. T., and McDonald, K. (2003). Cryoimmobilization and three-dimensional visualization of *C. elegans* ultrastructure. *J. Microscopy* **212,** 71–80.

Ohkura, K., Suzuki, N., Ishihara, T., and Katsura, T. (2003). SDF-9, a protein tyrosine phosphatase-like molecule, regulates the L3 dauer developmental decision through hormonal signaling in *C. elegans*. *Development* **130,** 3237–3248.

Paupard, M.-C., Miller, A., Grant, B., Hirsh, D., and Hall, D. H. (2001). Immuno-EM localization of GFP-tagged yolk proteins in *C. elegans* using microwave fixation. *J. Histochem. Cytochem.* **49,** 1–8.

Peckol, E. L., Zallen, J. A., Yarrow, J. C., and Bargmann, C. I. (1999). Sensory activity affects sensory axon development in *C. elegans*. *Development* **126,** 1891–1902.

Peckol, E. L., Troemel, E. R., and Bargmann, C. I. (2001). Sensory experience and sensory activity regulate chemosensory receptor gene expression in *Caenorhabditis elegans*. *PNAS* **98,** 11032–11038.

Perkins, L. A., Hedgecock, E. M., Thomson, J. N., and Culotti, J. G. (1986). Mutant sensory cilia in the nematode *C. elegans*. *Dev. Biol.* **117,** 456–487.

Peters, A., Palay, S. L., and Webster, H. de F. (1990). "The Fine Structure of the Nervous System." Oxford University Press, Oxford.

Phelan, P. (2000). Gap junction communication in invertebrates the innexin gene family. *Curr. Trends Membranes* **49,** 389–422.

Phelan, P., and Starich, T. A. (2001). Innexins get into the gap. *BioEssays* **23,** 388–396.

Raizen, D. M., and Avery, L. (1994). Electrical activity and behavior in the pharynx of *Caenorhabditis elegans*. *Neuron* **12,** 483–495.

Rappleye, C. A., Paredez, A. R., Smith, C. W., McDonald, K. L., and Aroian, R. V. (1999). The coronin-like protein POD-1 is required for anterior-posterior axis formation and cellular architecture in the nematode *Caenorhabditis elegans*. *Genes Dev.* **13,** 2838–2851.

Riddle, D. L. (1988). The Dauer Larva. *In* "The Nematode *Caenorhabditis elegans*." (W. B. Wood, Ed.), pp. 393–412. Cold Spring Harbor Laboratory Press, New York.

Rolls, M. M., Hall, D. H., Victor, M., Stelzer, E. H., and Rapaport, T. A. (2002). Targeting of rough endoplasmic reticulum membrane proteins and ribosomes in invertebrate neurons. *Mol. Biol. Cell* **13,** 1778–1791.

Rostaing, P., Weimer, R. M., Jorgensen, E. M., Triller, A., and Bessereau, J. L. (2004). Preservation of immunoreactivity and fine structure of adult *C. elegans* tissues using high-pressure freezing. *J. Histochem. Cytochem.* **52,** 1–12.

Ruvkun, G., and Hobert, O. (1998). The taxonomy of developmental control in *Caenorhabditis elegans*. *Science* **282,** 2033–2040.

Sagasti, A., Hobert, O., Troemmel, E. R., Ruvkun, G. B., and Bargmann, C. I. (1999). Alternative olfactory neuron fates are specified by the LIM homeobox gene *lim-4*. *Genes Dev.* **13,** 1794–1806.

Sato, M., Sato, K., Fonarev, P., Huang, C. J., Liou, W., and Grant, B. D. (2005). *Caenorhabditis elegans* RME-6 is a novel regulator of RAB-5 at the clathrin-coated pit. *Nature Cell Biol.* **7,** 559–569.

Sawin, E. R., Ranganathan, R., and Horvitz, H. R. (2000). *C. elegans* locomotory rate is modulated by the environment through a dopaminergic pathway and by experience through a serotonergic pathway. *Neuron* **26,** 619–631.

Schafer, J. C., Haycraft, C. J., Thomas, J. H., Yoder, B. K., and Swoboda, P. (2003). XBX-1 encodes a dynein light intermediate chain required for retrograde intraflagellar transport and cilia assembly in *Caenorhabditis elegans*. *Mol. Biol. Cell* **14,** 2057–2070.

Selkirk, M. E., Yazdanbaksh, M., Freedman, D., Blaxter, M. L., Cookson, E., Jenkins, R. E., and Williams, S. A. (1991). A proline-rich structural protein of the surface sheath of larval *Brugia* filarial nematode parasites. *J. Biol. Chem.* **266,** 11002–11008.

Sulston, J. E., Albertson, D. G., and Thomson, J. N. (1980). The *Caenorhabditis elegans* male: Postembryonic development of nongonadal structures. *Dev. Biol.* **78,** 542–576.

Sulston, J. E., and Horvitz, H. R. (1977). Post-embryonic cell lineages of the nematode, *Caenorhabditis elegans*. *Dev. Biol.* **56,** 110–156.

Sulston, J. E., Schierenberg, E., White, J. G., and Thomson, J. N. (1983). The embryonic cell lineage of the nematode *Caenorhabditis elegans*. *Dev. Biol.* **100,** 64–119.

Sulston, J. E., and Hodgkin, J. (1988). Methods. *In* "The Nematode *Caenorhabditis elegans*" (W. B. Wood, Ed.), pp. 587–606. Cold Spring Harbor Laboratory Press, New York.

Suzuki, H, Kerr, R., Bianchi, L., Froker-Jensen, C., Slone, D., Xue, J., Gerstbrein, B., Driscoll, M., and Schafer, W. R. (2003). *In vivo* imaging of *C. elegans* mechanosensory neurons demonstrates a specific role for the MEC-4 channel in the process of gentle touch sensation. *Neuron* **39,** 1005–1017.

Sym, M., Basson, M., and Johnson, C. D. (2000). A model for Niemann-Pick type C diseases in the nematode *Caenorhabditis elegans*. *Curr. Biol.* **10,** 527–530.

Thomas, C. F., and White, J. G. (1998). Four-dimensional imaging: The exploration of space and time. *Trends Biotechnol.* **16,** 175–182.

Thomas, J. H., Stern, M. J., and Horvitz, H. R. (1990). Cell interactions coordinate the development of the *C. elegans* egg-laying system. *Cell* **62,** 1041–1052.

Tobin, D. M., and Bargmann, C. I. (2004). Invertebrate nociception: Behaviors, neurons and molecules. *J. Neurobiol.* **61,** 161–174.

Troemel, E. R., Chou, J. H., Dwyer, N. D., Colbert, H. A., and Bargmann, C. I. (1995). Divergent seven transmembrane receptors are candidate chemosensory receptors in *C. elegans*. *Cell* **83,** 207–218.

Troemel, E. R., Kimmel, B. E., and Bargmann, C. I. (1997). Reprogramming chemotaxis responses: Sensory neurons define olfactory preferences in *C. elegans*. *Cell* **91,** 161–169.

Tsalik, E. L., Niacaris, T., Wenick, A. S., Pau, K., Avery, L., and Hobert, O. (2003). LIM homeobox gene-dependent expression of biogenic amine receptors in restricted regions of the *C. elegans* nervous system. *Dev. Biol.* **263,** 81–102.

Wadsworth, W. G., Bhatt, H., and Hedgecock, E. M. (1996). Neuroglia and pioneer neurons express UNC-6 to provide global and local netrin cues for guiding migrations in *C. elegans*. *Neuron* **16,** 35–46.

Wadsworth, W. G., and Hedgecock, E. M. (1992). Guidance of neuroblast migrations and axonal projections in *Caenorhabditis elegans*. *Curr. Opin. Neurobiol.* **2,** 36–41.

Walthall, W. W. (1990). Metamorphic-like changes in the nervous system of the nematode *Caenorhabditis elegans*. *J. Neurobiol.* **21,** 1085–1091.

Walthall, W. W., and Chalfie, M. (1988). Cell-cell interactions in the guidance of late-developing neurons in *C. elegans*. *Science* **239,** 643–645.

Ward, S., Thomson, J. N., White, J. G., and Brenner, S. (1975). Electron microscopical reconstruction of the anterior sensory anatomy of the nematode *C. elegans*. *J. Comp. Neurol.* **160,** 313–337.

Ware, R. W., and LoPresti, V. (1975). Three-dimensional reconstruction from serial sections. *Int. Rev. Cytology* **40,** 325–440.

Ware, R. W., Crossland, K., Russell, R. L., and Clark, D. V. (1975). The nerve ring of the nematode *C. elegans*: Sensory input and motor output. *J. Comp. Neurol.* **162,** 71–110.

Whittaker, A. J., and Sternberg, P. W. (2004). Sensory processing by neural circuits in *C. elegans*. *Curr. Opin. in Neurobiol.* **14,** 450–456.

White, J. G., Southgate, E., Thomson, J. N., and Brenner, S. (1976). Structure of the ventral nerve cord of *Caenorhabditis elegans*. *Phil. Trans. R. Soc. Lond. B* **275,** 327–348.

White, J. G., Albertson, D. G., and Anness, M. A. (1978). Connectivity changes in a class of motoneuron during the development of a nematode. *Nature* **271,** 764–766.

White, J. G., Southgate, E., Thomson, J. N., and Brenner, S. (1983). Factors that determine connectivity in the nervous system of *C. elegans*. *Cold Spring Harbor Symp. Quant. Biol.* **48,** 63–68.

White, J. G., and Horvitz, H. R. (1979). Laser microbeam techniques in biological research. Electro-Optical Systems Design, Aug. issue, 23–24.

White, J. G., Southgate, E., Thomson, J. N., and Brenner, S. (1986). The structure of the nervous system of *Caenorhabditis elegans*. *Phil. Trans. Roy. Soc., Series B Biol. Sci.* **314,** 1–340.

White, J. G., Southgate, E., and Thomson, J. N. (1992). Mutations in the *Caenorhabditis elegans unc-4* gene alter the synaptic input to ventral cord motor neurons. *Nature* **355,** 838–841.

Yeh, E., Kawano, T., Weimer, R. M., Bessereau, J.-L., and Zhen, M. (2005). Identification of genes involved in synaptogenesis using a fluorescent active zone marker in *C. elegans*. *J. Neurosci.* **25,** 3833–3841.

Yochem, J., Hall, D. H., Bell, L. R., Hedgecock, E. M., and Herman, R. K. (2005). Isopentenyl-diphosphate isomerase is essential for viability of *Caenorhabditis elegans*. *Mol. Genet. Genomics* **273,** 158–166.

Zhao, H. J., and Nonet, M. L. (2000). A retrograde signal is involved in activity-dependent remodeling at a *C. elegans* neuromuscular junction. *Development* **127,** 1253–1266.

Zhen, M., and Jin, Y. S. (1999). The liprin protein SYD-2 regulates the differentiation of presynaptic termini in *C elegans*. *Nature* **401,** 371–373.

Zhou, H. M., Brust-Mascher, I., and Scholey, J. M. (2001). Direct visualization of the movement of the monomeric axonal transport motor UNC-104 along neuronal processes in living *Caenorhabditis elegans*. *J. Neurosci.* **21,** 3749–3755.

INVESTIGATIONS OF LEARNING AND MEMORY IN *CAENORHABDITIS ELEGANS*

Andrew C. Giles, Jacqueline K. Rose, and Catharine H. Rankin

Department of Psychology and Brain Research Centre, University of British Columbia
Vancouver, British Columbia, Canada

Despite having a relatively small nervous system, *Caenorhabditis elegans* is capable of a great deal of behavioral plasticity. Each of the sensory systems that have been studied in *C. elegans* (mechanosensation, chemosensation, and thermosensation) has been shown to play a role in various forms of learning and memory. In this Chapter, we review work from each of the sensory systems and show simple forms of plasticity such as adaptation, non-associative learning, and different forms of classical conditioning. For each of these systems, the neural circuits have been identified and characterized. Taken together the results from these studies show that despite its small nervous system and its brief lifespan,

INTERNATIONAL REVIEW OF
NEUROBIOLOGY, VOL. 69
DOI: 10.1016/S0074-7742(05)69002-2

0074-7742/06 $35.00

C. elegans shows a remarkable capacity for plasticity. *C. elegans* is exquisitely tuned to gather information about experiences with its environment in order to find food and avoid areas without food. Genetic and pharmacologic analyses of these abilities as well as fluorescence microscopy of marker genes in identified neurons have begun to shed light on the molecular mechanisms of plasticity.

I. Introduction

Organisms must modify their behavior in response to experience in order to survive the constantly changing environment in which they live. Behavior is mediated by the nervous system, so for an organism to alter its behavior in response to experience with the environment, a mechanism for neural plasticity must exist. This process is known as learning. For learning to be advantageous, a process must also exist for storing modified behaviors so that during future interaction with the environment the learned behavior can be recalled. The storage and recall of this information is encompassed by the term "memory." From single-cell bacteria to highly complex mammals, all organisms have the capacity for both learning and memory. Behavioral characterizations of simple forms of learning and memory have shown that the rules for their acquisition and expression are remarkably similar across phylogeny, suggesting conservation of mechanisms through evolution. The molecular mechanisms that drive learning and memory are poorly understood in comparison.

Recently, *C. elegans* has been identified as a useful model for studying learning and memory at the cellular and molecular level for a number of reasons: (1) the vast array of relatively simple, well-characterized behaviors observed in *C. elegans*; (2) the knowledge of worm genetics and ease with which genetic manipulation can be performed; (3) the simplicity of the *C. elegans* nervous system, which contains only 302 neurons, which are all identified and for which the connectivity is known; and (4) the *in vivo* whole-organism molecular analyses that are available with the recent advances in microscopy and genetic-based fluorescence imaging. Before discussing the ways in which learning and memory are studied in *C. elegans* and the contributions that these studies have made, some general principles concerning learning and memory will be outlined.

Learning and memory are not directly measurable; they must be indirectly measured by performance. For this reason, controls are very important in order to rule out other factors that can affect performance. These include deficits in sensory and motor systems, which could be the unexpected result from a genetic manipulation or injury from physical damage by handling or by a pharmacologic treatment. Therefore, learning and memory must always be inferred by significant changes in spontaneous or stimulus-evoked behavior measured before, during, or after a training (learning) procedure. This is an important concept to keep in mind as the learning and memory studies are discussed.

The simplest forms of learning exhibited by organisms are non-associative. Non-associative learning is a change in the attention that an organism directs to an individual stimulus. This includes habituation and sensitization. Habituation is a decrease in attention towards an irrelevant stimulus, while sensitization is an increase towards a potentially significant stimulus. Both are measured by the change in the behavioral response to the given stimulus; therefore, habituation is observed when a response decreases with multiple stimulation, while sensitization is measured by an increase in response to a stimulus. Groves and Thompson (1970) reviewed these two contrasting behaviors theorizing that they are both part of a dual-process mechanism that constantly weighs the contributions of both habituation and sensitization in order to yield a final behavioral response. This theory will be discussed in relation to the findings from studies of learning and memory in *C. elegans.*

More complex learning includes associative forms, such as classical conditioning. Classical conditioning involves the association of a neutral stimulus, known as the conditioned stimulus (CS), with a biologically relevant stimulus that evokes a response prior to training (unconditioned stimulus [US]). The naturally occurring response to the US is termed the unconditioned response (UR). When the CS has been repeatedly paired with the US and an association has been formed, the presentation of the CS alone evokes a similar response as the UR, called the conditioned response (CR). Several forms of classical conditioning have been observed in *C. elegans*, which will be discussed.

Memory, which is evident with all these different forms of learning, is usually discussed at three temporal levels—short-term, intermediate-term, and long-term; however, intermediate-term memory is not yet well-described in *C. elegans* and will not be discussed in this chapter. Unfortunately, it is unclear how time scales relate across different organisms that have drastically different lifespans, but it is generally thought that short-term memory is based on the level of performance immediately after training, while long-term memory is based on performance 24 h or so thereafter. Both forms are observed in *C. elegans* and will be compared and discussed in further detail.

In reviewing the study of learning and memory in *C. elegans*, this chapter has been divided into three sections which correspond to the different sensory modalities that transduce the stimuli for the various learned responses: mechanosensation, chemosensation, and thermosensation. Most of the research focuses on learning involving a single-sensory system; however, recent studies investigate the effects of multi-modal paradigms, such as context conditioning of the tap withdrawal response (Rankin, 2000), which uses both mechanical and chemical stimuli. The discussion will begin with mechanosensation because learning in

C. elegans was first described as such using the tap withdrawal response (Rankin *et al.*, 1990).

II. Mechanosensory Learning and Memory

Mechanosensation refers to the ability to sense tactile and vibrational stimuli. In *C. elegans* this includes touches to the body as well as vibrations in the substrate produced by a tap to the surrounding environment, usually a plastic agar filled petri dish in the laboratory. Briefly, *C. elegans* possess a number of mechanosensory neurons, including the ALM, AVM, and PLM touch sensory neurons. The touch sensory neurons are important in mediating head touch (ALM and AVM), tail touch (PLM), and the tap withdrawal response (ALM, AVM, and PLM). An advantage to using tap withdrawal is that the response is produced by delivering a mechanical stimulus that is constant in intensity (unlike touch which can differ in both intensity and in placement of stimulation from a handheld stimulus such as an eyelash mounted on a syringe). These touch sensory neurons synapse on to four pairs of interneurons—AVD, AVB, AVA, and PVC (often called the command interneurons)—which integrate information and synapse onto relevant motor neurons (for neural circuit see Fig. 1). The tap withdrawal response occurs in response to a tap to the side of the plastic petri plate holding the agar and the worm. In response to the tap, a worm that is swimming forward reverses direction, swimming backwards for a brief period after which it swims forward again (Rankin *et al.*, 1990; Chiba and Rankin, 1990).

A. Habituation

In 1990, Rankin *et al.* were the first to characterize learning in *C. elegans.* They measured the length of the path along which a worm reversed after a train of mechanical taps (7 taps in 600 ms), and found that the distance decreased after repeating the stimulus 40 times at regular interstimulus intervals of 10 s. This response decrement recovered back to baseline within a few minutes (spontaneous recovery). In order to determine whether the decrement in responding was due to habituation, sensory adaptation, or fatigue, this behavior must show dishabituation. Rankin *et al.* (1990) used an electric shock administered through the agar on either side of the worm as a dishabituating stimulus. Following the shock, the previously habituated worm responded to the tap stimulus with large reversals, indicating that the decrease in responding observed was in fact habituation.

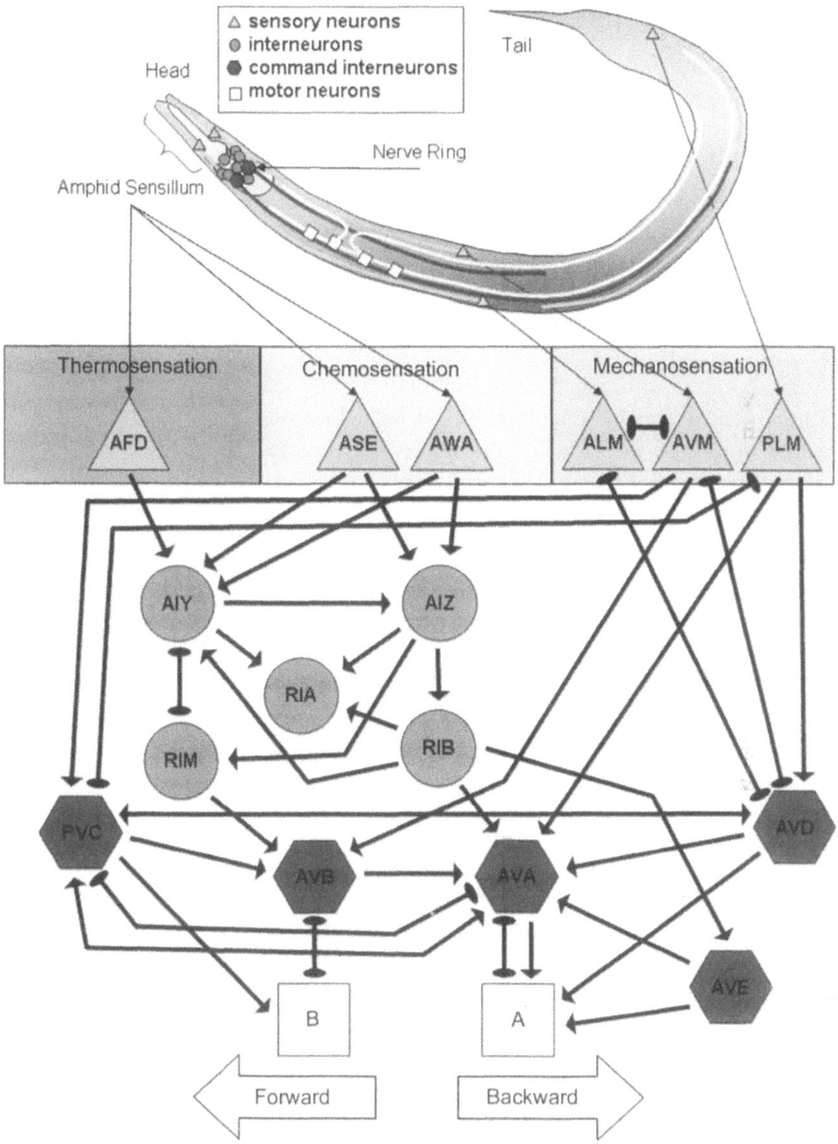

FIG. 1. Simplified and integrated schematic diagram of the major elements of the neural circuits involved in mechanosensory, chemosensory, and thermosensory behaviors that have been used to demonstrate learning and memory and their anatomical locations within the worm. Green triangles represent sensory neurons, red circles represent interneurons, blue hexagons represent command interneurons, and yellow squares represent pools of motor neurons. Arrows represent chemical synaptic connections while bars represent gap junctions. Based on the hypothesized neural circuits created by Chalfie *et al.* (1985), Wicks and Rankin (1995), Mori and Oshima (1995), and Hobert (2003). (See Color Insert.)

Rankin and Broster (1992) investigated the factors that influence short-term habituation and spontaneous recovery from habituation of the tap withdrawal response. In the first experiment, they investigated the effects of different interstimulus intervals (ISIs) on habituation. Worms were given 30 stimuli at 2, 10, 30, and 60 s ISIs. The results showed that after worms were trained at shorter ISIs (i.e., 10 s) they produced small responses that were $<10\%$ of their initial response magnitude, while worms trained at longer ISIs (i.e., 60 s) produced intermediate responses that were approximately 30% to 40% of their initial response magnitude; thus, short ISIs produced a deeper level of habituation than long ISIs. Spontaneous recovery from habituation was also investigated by measuring the tap withdrawal response to a single train of taps 30 s, 10 min, 20 min, and 30 min after habituation training. Worms habituated using shorter ISIs recovered more rapidly than worms habituated at longer ISIs. This was an important result as it further supported the idea that the observed decrement in reversals was in fact habituation and not fatigue because worms trained at short ISIs, who showed a deeper level of habituation, actually recovered more quickly than worms trained with longer ISIs. If the worms were fatigued, the opposite would be expected. The differences observed in spontaneous recovery might have been the result of the different levels of habituation produced by short and long ISIs. Rankin and Broster (1992) addressed this by training worms with short and long ISIs to the same predetermined level of habituation. This allowed spontaneous recovery from habituation with different ISIs to be compared from comparable levels of habituation. Worms trained with this protocol still showed rapid recovery following short ISIs and slower recovery following long ISIs, supporting the hypothesis that ISI determines the rate of spontaneous recovery and not the level of habituation.

The fact that level of habituation is not a factor of recovery was also supported by an elegant study by Wicks and Rankin (1996). Previously, the neural circuit of the tap withdrawal response had been identified (Wicks and Rankin, 1995). Three neurons (ALM, AVM, and PLM) are important for the sensory input of the tap withdrawal response (Fig. 1); by ablating subsets of these neurons using a laser, the amount of sensory input could be modulated. Wicks and Rankin (1996) found that worms with PLM ablations habituated to a train of taps at 10 s ISI with similar kinetics and to a similar level of habituation as non-ablated worms trained at a 60 s ISI. However, the recovery of the PLM ablated worms mimicked the recovery of 10 s ISI non-ablated worms, further supporting the idea that spontaneous recovery is an ISI-dependent process.

The aforementioned experiments investigated the effect of the kinetics and level of habituation on spontaneous recovery after habituation had reached an asymptotic level; however, the effect of habituation before reaching an asymptotic level was still to be addressed. To investigate this, Rankin and Broster (1992) trained groups of worms at 10 and 60 s ISIs with trains of taps using only a very small number of stimuli (eight), so that worms did not reach an asymptotic level. After determining the mean response, each group was split into two groups that had either habituated to levels lower or higher than the overall mean. They found that the worms who had habituated to a level lower than the mean showed the characteristic recovery curve for that ISI, while the group that habituated to levels above the mean recovered much more quickly. This demonstrated that once worms reach an asymptotic level, the rate of recovery is determined by the ISI of training.

Finally, Broster and Rankin (1994) investigated the effect of changing from one ISI to another in the middle of training. Worms who were initially trained at a 10 s ISI were switched to a 60 s ISI schedule and exhibited a level of habituation and a rate of spontaneous recovery similar to worms trained at a 60 s ISI for the entire training session. Similarly, worms initially trained at a 60 s ISI and switched to a 10 s ISI showed a level of habituation and recovery similar to worms trained at a 10 s ISI for the whole period. Thus, both the level of habituation and recovery from habituation depend on the ISI of the second half of the training. Rankin and Broster (1992), Broster and Rankin (1994), and Wicks and Rankin (1996) characterized the non-associative learning observed during habituation of the tap withdrawal response. They found that habituation to tap was dependent on the ISI of the training schedule and the level of habituation before, but not after reaching the asymptotic level of the response. As well, they discovered that the level of habituation and the rate of spontaneous recovery were dependent on the most recently experienced schedule.

The effects of life-span development on short-term habituation to the tap withdrawal response as well as the behavior itself have also been investigated (Beck and Rankin, 1993; Chiba and Rankin, 1990; Rankin *et al.*, 2000). Beck and Rankin (1993) initially studied the effects of aging, by comparing habituation in 4-, 7-, and 12-day-old worms (average lifespan of a worm is 14 days). Four-day old worms are at their reproductive prime and are considered to be adults. Despite the fact that 12-day-old worms had significantly smaller reversal magnitudes compared to the 4- and 7-day-old worms, all age groups showed similar 10 s ISI habituation. Further, both 7- and 12-day-old worms showed a deeper level of habituation than 4-day-old worms during the 60 s ISI training suggesting older worms might be more susceptible to fatigue. This was supported by the observations that these older worms did not show dishabituation to the same extent as younger worms, and that 12-day-old worms recovered more slowly from habituation than younger worms. These data showed that worms habituate more rapidly and recover more slowly as they age.

Rankin *et al.* (2000) also investigated habituation in early development. This was challenging because the actual tap withdrawal response changes over the course of larval development (Chiba and Rankin, 1990). Initially, larval worms accelerate approximately 50% of the time and reverse the other half of the time in response to the mechanical tap, while adult worms almost always reverse. This switch reflects the development of the neural components of the circuit (Fig. 1); in larval animals only the ALM pair of head neurons are present, but by the end of larval stage 4 the third head neuron, AVM, has grown in and made its connections (Chalfie *et al.*, 1985). Backward movement is mediated through activation of the AVM- and ALM-driven neural circuit, while forward movement is mediated by the PVM driven circuit (Chalfie *et al.*, 1985; Wicks and Rankin, 1995) (Fig. 1). Because the integration of these two circuits changes over the course of development, each circuit was investigated separately (Rankin *et al.*, 2000) by laser ablating ALM to analyze the habituation of forward responses to tap mediated by PLM, and ablating PLM to study the habituation of backward movement mediated by ALM. Rankin *et al.* (2000) looked at habituation to short and long ISIs at six different developmental ages; each of the larval stages (L1 to L4), young adults (3-day-old), and adults (4-day-old) were examined under either ALM- or PLM-ablated conditions. They found that for both forward and backward circuits, all age groups showed similar habituation and recovery curves at 10 and 60 s ISI as observed in adult worms in previous studies (Broster and Rankin, 1994; Rankin and Broster, 1992). These data show that the characteristics of habituation that are present immediately after hatching remain consistent during all developmental stages through to adulthood.

B. Neural Circuit Analyses

Thus far, the discussion has focused on the characterization of the behavioral aspects of short-term habituation to tap. The mechanism that mediates this habituation has also been investigated. The neural circuit of the tap withdrawal response significantly overlaps with the neural circuit for spontaneous reversals and for the thermal avoidance response, sharing most of the same interneurons and motor neurons that compose the locomotion pattern generator (Fig. 1). The difference between these circuits is the source of the sensory input; ALM, AVM, and PLM are the sensory neurons required for the tap withdrawal response (Wicks and Rankin, 1995), while AFD is the sensory neuron for thermal stimuli (Mori and Oshima, 1995). There is no obvious sensory input mediating spontaneous reversals. Wicks and Rankin (1995) hypothesized that if the locus of plasticity were in the interneurons or motor neurons, then habituating a worm to tap would affect its ability to respond to other stimuli, for example, heat. Wicks and Rankin (1997) showed that habituation to tap had no effect on the magnitude or frequency of either spontaneous reversals or the thermal avoidance response, suggesting that the locus of plasticity is situated presynaptic to the command interneurons of the locomotion pattern generator. Therefore, the most likely sites for plasticity are the synapses between the mechanosensory neurons (ALM, AVM, and PLM) and the interneurons.

C. Genes That Affect Habituation

Rankin and Wicks (2000) extended these findings by examining the role of a gene related to neurotransmission that is expressed in the sensory neurons. Previous studies suggested that the mechanosensory neurons release glutamate. The *C. elegans* homologue of a mammalian glutamate vesicular transporter (VGLUT1), known as EAT-4 (Bellocchio *et al.*, 2000), is expressed in the touch sensory neurons (Lee *et al.*, 1999); GLR-1 and NMR-1, which are *C. elegans* homologues of mammalian AMPA/Kainate-type and NMDA glutamatergic receptors, respectively, are expressed by AVA, AVB, AVD, and PVC, the interneurons of the tap withdrawal response circuit (Brockie *et al.*, 2001; Hart *et al.*, 1995; Maricq *et al.*, 1995). Rankin and Wicks (2000) pursued the hypothesis that if the synapse between the mechanosensory neurons and the interneurons was the site of plasticity, then worms deficient in glutamate transmission would show abnormalities in their habituation to tap. This was investigated by studying the mutant strain *eat-4*. Rankin and Wicks (2000) found that the *eat-4* mutant strain shows normal tap withdrawal responses, but habituated significantly more rapidly and to a deeper habituation level than wild-type worms. Although this effect occurred in both of the ISIs, worms still habituated to the 10 s ISI more rapidly and more completely than worms habituated at 60 s ISI. By 10 min after habituation training, *eat-4* mutants had not recovered as quickly as wild-type worms, even though recovery still occurred in an ISI-dependent fashion. Interestingly, *eat-4* worms did not show dishabituation after receiving shock following habituation suggesting that the EAT-4 glutamate vesicular transporter is a critical component in the mechanism of habituation.

A second neurotransmitter has been identified that affects short-term habituation to tap. The role of dopamine, which is known to be a neural modulator in mammalian nervous systems, was briefly studied by Sanyal *et al.* (2004). They found that both *cat-2* mutants, which lack the tyrosine hydroxylase enzyme (necessary for the production of dopamine), and *dop-1* mutants, which do not express a dopamine receptor normally expressed on the mechanosensory neurons, habituated more quickly and more deeply than wild-type as measured by number of animals reversing to each tap. Unfortunately, when reversal length was measured during habituation the *dop-1* mutants were not different from wild-type; however, these results suggest a possible role of dopamine in modulating habituation to tap.

Further genetic dissection of short-term habituation of the tap withdrawal response was conducted by Xu *et al.* (2002), who isolated a mutant named *hab-1* from a genetic screen. The *hab-1* worms habituated more slowly than wild-type worms despite the fact that these mutants showed similar responses to single tap and wild-type dishabituation. This is an interesting mutant because it has the opposite phenotype to the glutamate and dopamine mutants, which both showed quicker habituation than wild-type worms. Unfortunately, the gene product of *hab-1* has not been identified, but it appears to play a role in the mechanism underlying habituation to tap.

D. Long-Term Memory for Habituation

The discussion so far has been focused on addressing the behavioral characteristics and molecular mechanisms involved in the acquisition and short-term memory of non-associative habituation to a mechanosensory response. Long-term memory for habituation is also of great interest. Long-term memory for habituation of the tap withdrawal response was first reported in Rankin *et al.* (1990). They demonstrated 24-h retention of habituation of tap withdrawal using a modified *Aplysia* long-term memory protocol (Carew *et al.*, 1972), consisting of five blocks of training, each separated by a 1-h resting period, followed by a single test block 24 h later. The experimental group showed a significantly smaller mean reversal length from the first training block compared to the mean reversal length from the testing block 24 h later; no difference was observed in the control group.

Beck and Rankin (1997) examined the factors that affect long-term memory for habituation in *C. elegans*, by further characterizing the training parameters necessary for memory formation through examination of the ways that the number of stimuli, ISI, and type of training affected long-term memory in the worm. Two types of training were compared: distributed (spaced) training, which features the division of all stimuli into blocks separated by hour-long rest periods; and massed training, where all of the training stimuli are delivered in one single block. The results indicated that distributed training at a 60 s ISI with either 40 (4 blocks of 10 taps) or 60 (3 blocks of 20 taps) stimuli produced significant long-term memory. In general, more stimulation (60 stimuli) was better at producing long-term memory than less stimulation (40 stimuli). Training with a 10 s ISI did not reliably produce long-term memory. In summary, long-term memory of habituation was most reliably produced by training protocols with a high total number of stimuli, long ISIs, and distributed training. This superiority of distributed or spaced training over massed training appears to be a fundamental feature of memory; for example, it has been shown in long-term habituation in *Aplysia* (Carew *et al.*, 1972), in classical conditioning in *Drosophila* (Tully *et al.*, 1994), and in memory for lists of nonsense syllables in humans (Ebbinghaus, 1885).

Both vertebrate and invertebrate studies have shown that protein synthesis is necessary for the formation of long-term memory (Davis and Squire, 1984; Flood *et al.*, 1977; Frey *et al.*, 1988; Ghirardi *et al.*, 1995; Mirumori *et al.*, 1985; Montarolo *et al.*, 1986; Tully *et al.*, 1994). Beck and Rankin (1995) examined the effect of protein synthesis inhibition on long-term memory of habituation in *C. elegans* using heat shock. Forms of cellular stress, such as heat shock, cause all basal protein synthesis to stop in order for the rapid synthesis of a class of heat shock proteins important for dealing with stress. An advantage to using heat shock (rather than pharmacologic blockade) to inhibit protein synthesis is the discrete temporal control that can be used for its application.

Using heat shock (45 min at a temperature of 32° C), Beck and Rankin (1995) investigated the effects of heat shock on memory when applied at three different times: prior to training, during training (administered during each of the hour-long rest periods), and prior to testing. In all cases, heat shock had no effect on the tap withdrawal response itself or the short-term kinetics and level of habituation observed during training. Heat shock treatment, delivered during training, blocked significant decrease of the mean reversal length during testing compared to the mean reversal length from the first training block. In a subsequent experiment, Beck and Rankin (1995) found that heat shock treatment is more effective if delivered in the first 30 min of the 1-h rest periods compared to later in the rest period. Heat shock prior to training or prior to testing had no effect suggesting that protein synthesis inhibition between the training blocks attenuates formation of long-term memory of habituation in *C. elegans*.

The experiments by Beck and Rankin (1995, 1997) indicated that long-term memory could be studied in *C. elegans*; however, the protocol used in these experiments was very time intensive requiring 5 h of training for each worm. At this point a more efficient and easier procedure for producing long-term memory of habituation was developed by Rose *et al.* (2002). The original procedure compared reversal responses made during the first training block to those made in the test block. The new procedure compared responses of a group of trained worms to the responses of a group of untrained controls. Worms received training (4 blocks of 20 taps at a 60 s ISI with 1 h rest periods between blocks) grouped together on a single plate while the matched controls were given a single tap on training day. Twenty-four hours after training, individual worms from both trained and control groups were given a single test block of taps at a 60 s ISI. In this way reversal responses made during the test blocks for trained worms could be compared to control worms (Fig. 2). Using this procedure, long-term memory for habituation results in significantly smaller mean reversal lengths for the 10 test taps from trained worms compared to control worms. Massed training (a single block of 80 taps at a 60 s ISI) resulted in no difference in reversal magnitudes between trained and control worms, supporting the fact that massed training does not produce long-term memory. As well, distributed training at a 10 s ISI again produced no memory for habituation. Finally, the sensitivity of long-term memory of habituation to heat shock delivered between training blocks

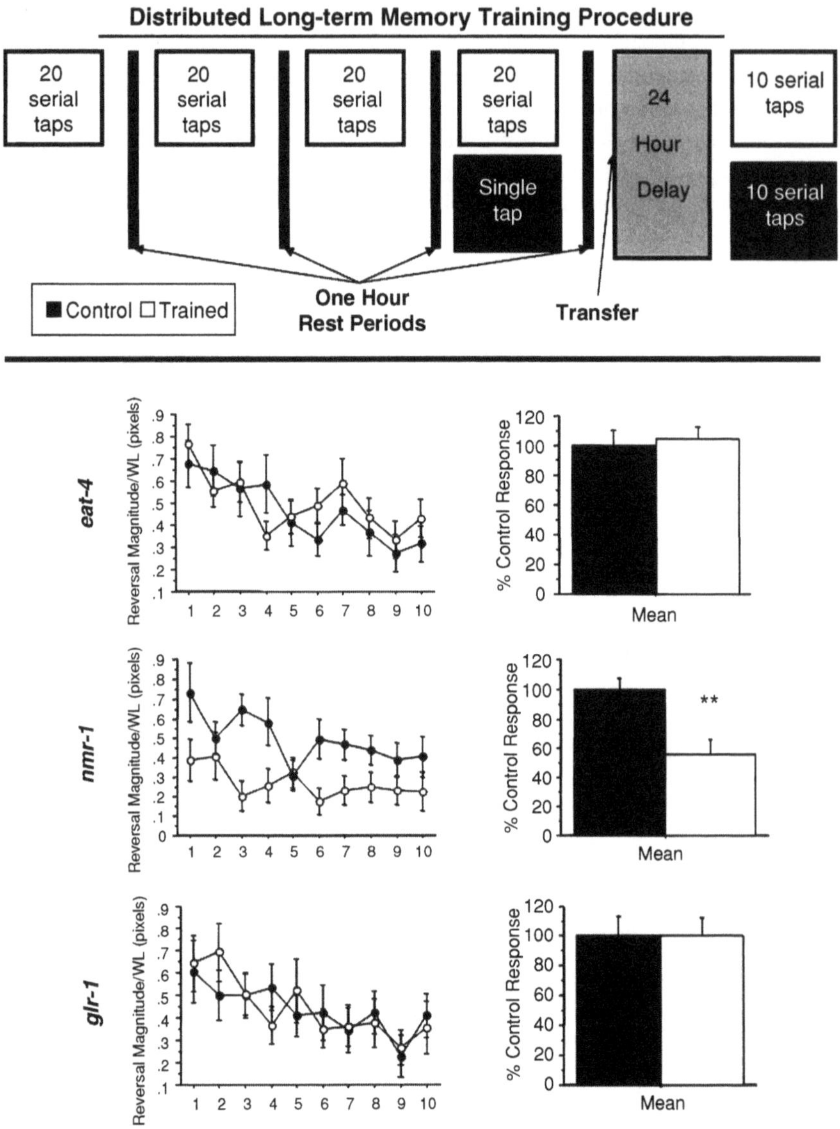

FIG. 2. Schematic diagram of the long-term memory distributed training paradigm (Top). Mean ± SEM of the reversal magnitude/worm length (WL) during the 10 test taps 24 hours after training for trained and control *eat-4, nmr-1,* and *glr-1* mutant strains (Bottom left). Mean ± SEM of the mean reversal magnitude/WL during the 10 test taps 24 hours after training expressed as a percentage of controls for trained and control, *eat-4, nmr-1,* and *glr-1* mutant strains (bottom right). **Significant difference ($p < 0.01$).

was replicated suggesting that the memory reported with this new training protocol was characteristically the same as that studied previously by Beck and Rankin (1995, 1997). Cold shock had previously been used as a method of blocking protein synthesis in *Drosophila* (Tully and Quinn, 1985); however, it does not appear to block memory for long-term habituation in *C. elegans* (J. Rose, 2002, unpublished data). These studies demonstrated that this new procedure produced reliable long-term memory and could be used as a faster and easier way to screen mutant strains for deficits in long-term memory of habituation.

E. Glutamate Neurotransmission and Memory

Using this new protocol, Rose *et al.* (2002) began examining the molecular mechanisms of long-term memory by studying the *eat-4* worms that lacked the glutamate vesicular transporter. As previously discussed, *eat-4* mutants show enhanced short-term habituation to tap (Rankin and Wicks, 2000) suggesting that the *eat-4* mutants might also show enhanced memory. Surprisingly, Rose *et al.* (2002) found that trained *eat-4* worms did not respond differently than control *eat-4* worms (Fig. 2). The DA2142 transgenic rescue strain, which was used by Rankin and Wicks (2000) to save abnormal habituation, rescued long-term memory; trained DA2142 worms showed significantly smaller reversals than control DA2142 worms. This suggests that presynaptic glutamate release, deficient in *eat-4* worms, is essential for the formation of long-term memory.

The data from the *eat-4* experiment suggested that glutamate mediates long-term memory of habituation. The next logical step was to investigate the role of glutamatergic receptors expressed on the interneurons of the tap circuit in long-term memory. The command interneurons express at least two different glutamate receptors, GLR-1, which is a homologue of the mammalian Kainate/AMPA receptor type (Hart *et al.*, 1995; Maricq *et al.*, 1995) and *nmr-1* which is a homologue of the mammalian NMDA-type glutamate receptor (Brockie *et al.*, 2001). In mammals both of these receptors play a role in synaptic plasticity (for review, see Lüscher and Frerking, 2001; Malinow and Malenka, 2002). *nmr-1* worms showed normal long-term memory for habituation (J. Rose, 2003, unpublished data) (Fig. 2). However, when *glr-1* mutant worms were tested for long-term memory of habituation, trained worms were not significantly different from control worms (Rose *et al.*, 2003) (Fig. 2), suggesting that activation of GLR-1 receptors is necessary for memory. This hypothesis was supported by pharmacologic experiments that found that worms trained in the presence of DNQX (a non–NMDA-type glutamate receptor antagonist) did not show long-term memory and did not respond differently to control worms (Rose *et al.*, 2003).

In mammals, increases or decreases in synaptic strength are often associated with changes in the expression of glutamate receptors (Lüscher and Frerking,

2001; Malinow and Malenka, 2002). Rose *et al.* (2003) investigated whether long-term memory training for habituation was accompanied by a change in the expression of GLR-1 using a strain of worms expressing GLR-1 fused to green fluorescent protein (GFP). The amount of GLR-1:GFP expressed by a worm is quantifiable using images taken using a confocal microscope. When worms carrying a GLR-1:GFP transgene were given distributed long-term habituation training, expression of GLR-1:GFP in trained worms was found to be significantly lower than GLR-1:GFP expression in controls. Interestingly, there was no difference in the number of GLR-1:GFP expressing clusters. Rose *et al.* (2003) hypothesized that the number of receptors at each synapse had decreased as a result of memory training, but that the number of synapses had not changed. These results suggest that long-term memory for habituation is molecularly mediated by a down-regulation of GLR-1 expression. Together, these studies suggest an important role for glutamate neurotransmission, and more specifically glutamate receptor trafficking, in the formation of long-term memory of habituation of the tap withdrawal response. This mechanism is remarkably similar to the mechanisms described for some forms of long-term depression in mammals.

F. Context Conditioning and Habituation

Non-associative learning and memory have traditionally been thought to be completely dependent on the parameters of a single stimulus (Groves and Thompson, 1970). A number of studies involving a variety of organisms have shown that although initial non-associative learning is exclusively dependent on this one stimulus, future responses to the presentation of this stimulus can be dependent on associations made with other stimuli during the initial training. Wagner (1976, 1978, 1979) originally made this hypothesis suggesting that habituation is non-associative, but memories can be formed from previous habituation experiences and these memories can include associations. In particular, the context or environmental cues that were present during training can be associated with the training; the process of this association is known as context conditioning. The formation of context associations with habituation have been shown in a number of studies, suggesting that context is important in mediating memory for habituation (Evans and Hammond, 1983).

Rankin (2000) first demonstrated context conditioning in *C. elegans* by exploring the effects of a chemosensory context on the memory for habituation of the tap withdrawal response. The context cue used was a concentration of 0.4 M of Na^+ ions; this same chemical cue had been shown to produce classical conditioning for chemotaxis (Wen *et al.*, 1997; discussed in later text). The Na^+ ions that served as the context cue did not affect the response to tap nor the level of habituation to tap, so it was considered an appropriate stimulus for context. Worms were habituated on either plain agar or Na^+ treated agar with 30 taps at either a 10 or 60 s ISI. During a 1-h rest period, worms were transferred onto fresh plain agar plates; at the end of the rest period worms were transferred and re-habituated (tested) on either a plain agar or Na^+ agar plate. Worms trained and tested on the same type of agar plate (Na^+) showed greater retention of the earlier training shown by decreased responses during the second round of habituation compared to the original training, suggesting worms associated the Na^+ context with tap withdrawal habituation. Worms trained and tested on different types of agar plates did not show a significant decrease in mean reversal lengths between the two sessions. In order for context conditioning to occur the environment had to be paired with the taps; several experiments were run in which worms experienced the Na^+ environment without taps. When worms trained and tested on Na^+ agar plates were left on the Na^+ context for the hour between training and testing, the context effect was lost; this is extinction. In addition, if worms were placed on the Na^+ agar context for 1 h before the initial training, the context effect was lost; this is latent inhibition or CS pre-exposure. Together, the extinction and latent inhibition experiments support the notion that context conditioning was a form of associative conditioning.

G. Summary of Mechanosensory Learning and Memory

To briefly review mechanosensory learning and memory, Rankin *et al.* (1990) were the first to show learning and memory in *C. elegans* by studying habituation, a non-associative form of learning, of the response to a mechanosensory stimulus. Since then, developmental and behavioral factors that affect habituation were identified. Of these factors, the ISI of the training schedule seems to be the most important, affecting the kinetics and level of habituation, the rate of spontaneous recovery after habituation, and the formation of long-term memory. Neural circuit analyses led to the hypothesis that the site of plasticity is located at the glutamatergic synapse between the mechanosensory neurons and command interneurons. More recently, the cellular and molecular mechanisms that mediate the behavior have been investigated. Pre-synaptic release of glutamate from the sensory neurons is essential for wild-type short-term habituation; pre-synaptic glutamate release, post-synaptic glutamate receptor activation, and protein synthesis are all necessary for the formation of long-term memory of habituation to the tap withdrawal response. Long-term habituation of the tap response correlates with a down-regulation of the expression of GFP-tagged GLR-1 receptors. Finally, the memory of previously experienced habituation to mechanosensory stimuli can be associated with chemosensory cues using context conditioning, supporting the idea that *C. elegans* is capable of more complex forms of learning.

This idea will be further explored in the chemosensory and thermosensory learning and memory sections that follow.

III. Chemosensory Learning and Memory

Chemotaxis is characterized by the tendency of worms to migrate towards or away from certain chemicals and substrates. Briefly, *C. elegans* will approach cations (e.g., Na^+, Mg^{2+}, K^+, and so on), anions (e.g., Cl^-), cyclic nucleotides (e.g., cAMP and cGMP; Ward, 1973) as well as volatile chemicals such as benzaldehyde, diacetyl, and isoamyl alcohol (Bargmann and Mori, 1997). Traditionally, there have been two measures of chemotaxis: one is orientation, observed by tracking the path of individual worms as they move towards or away from a chemical point source; and two is accumulation, measured from a population of worms as the number of worms that gather at the chemical location. It has been found that this chemical-orienting behavior is also subject to plasticity.

Eleven pairs of amphid neurons (found at the nose of the worm) are responsible for chemotaxis, eight of which have sensory endings that contact the external environment directly and detect mostly water-soluble attractants, while the remaining three chemosensory amphids are enclosed by the amphid sheath cell and detect mostly volatile chemicals. To be able to respond to so many different substances, individual chemosensory amphid neurons are likely responsible for the detection of multiple chemicals. This ability for a single neuron to detect multiple chemicals suggests that in some instances, chemotaxis plasticity may rely on an intraneuronal site of plasticity.

A. Adaptation

The first evidence of neural plasticity within the chemosensory system was olfactory adaptation. Colbert and Bargmann (1995) developed a paradigm where worms were exposed to a high but uniform concentration of a given odorant for a measured period, referred to as the adaptation period. After this exposure, the worms were transferred to a chemotaxis assay plate, which contained the test odorant on one side and a control substance on the opposite side (for summary of procedure, see Fig. 3). They found that the chemotaxis index ([number migrated to test − number migrated to control]/total number) for a number of attractant odorants, such as benzaldehyde and diacetyl, decreased as the duration of the adaptation period increased. This process was reversed if worms experienced

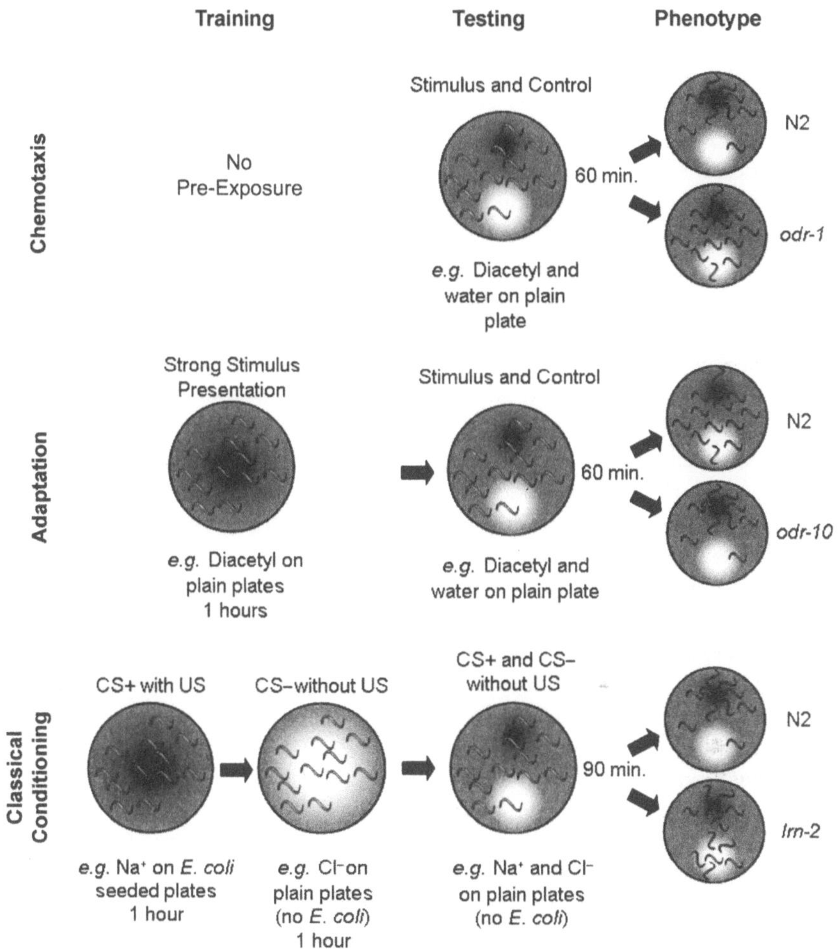

FIG. 3. Representation of the chemotaxis, olfactory adaptation, and chemosensory classical conditioning paradigms. Phenotypes of N2 (wild-type), *odr-1, odr-10,* and *lrn-2* strains are represented in the far right column. N2 worms are attracted to diacetyl during the chemotaxis assay, less attracted towards diacetyl during the adaptation assay, and attracted to Na^+ (CS+) during the classical conditioning assay (Bernhard and van der Kooy, 2000; Wen *et al.*, 1997). *odr-1* mutants are not attracted to diacetyl during the chemotaxis assay (Bernhard and van der Kooy, 2000). *odr-10* mutants do not show decreased attraction to diacetyl during the adaptation assay (Bernhard and van der Kooy, 2000). *lrn-2* mutants show no preference for either the Na^+ (CS+) or the Cl^- (CS−) during the classical conditioning assay (Wen *et al.*, 1997).

some time (e.g., 3 h) free from the given odorant. Interestingly, adaptation to one odorant does not always carry over to all other odorants sensed by that same neuron. For example, benzaldehyde, butanone, and isoamyl alcohol all activate AWC (Bargmann *et al.*, 1993). When a worm was adapted to benzaldehyde, cross-adaptation to isoamyl alcohol was also noted; however, adaptation to either benzaldehyde or isoamyl alcohol did not produce cross-adaptation with butanone (Colbert and Bargmann, 1995).

However, there are additional complications in understanding the mechanisms responsible for adaptation. For instance, adaptation to benzaldehyde and butanone, two substances that show no cross-adaptation, both rely on calcium influx; worms pre-exposed to the calcium chelator EGTA showed a deficit in adaptation to benzaldehyde and butanone, but low extracellular calcium appeared to have no effect on adaptation to isoamyl alcohol (AWC) or diacetyl (a substance that activates the AWA neuron) (Colbert and Bargmann, 1995).

Colbert and Bargmann (1995) also performed a genetic screen to isolate mutant strains that displayed defective adaptation and discovered *adp-1*, which showed deficits in olfactory adaptation to benzaldehyde and butanone. A second mutant, *osm-9*, defective at avoiding highly osmotic concentrations, also showed defective adaptation to butanone and isoamyl alcohol. Neither of these mutants was characterized as having chemosensory deficits because they behaved like wild-type during unadapted chemotaxis assays. Interestingly, the deficits caused by *adp-1* mutation and EGTA treatment were not additive suggesting that *adp-1* may play a role in intracellular calcium signaling in the sensory neurons (Colbert and Bargmann, 1995). In a later report, Colbert *et al.* (1997) found that the molecular product of the *osm-9* gene is a transient receptor potential (TRP)-like channel. Because of the behavioral deficiency observed in the mutant strain, this TRP-like channel is implicated in the molecular mechanism of adaptation (Colbert *et al.*, 1997). Interestingly, these TRP-like channels are important for the activation of G-protein-coupled odor receptors, which have recently been identified to be the gene product of *odr-10* (Sengupta *et al.*, 1996; discussed in the following text). These studies provide a number of clues towards uncovering the molecular mechanism responsible for mediating chemosensory adaptation.

Two other genes (*odr-1* and *odr-10*) have been studied for their roles in chemosensory adaptation, which have previously been identified using a chemotaxis assay (Bargmann *et al.*, 1993; Sengupta *et al.*, 1996). Bernhard and van der Kooy (2000) examined the effect of mutations in these genes using a chemotaxis assay that measured adaptation to diacetyl and found that *odr-10* performs like wild-type during naive chemotaxis but shows no adaptation, while *odr-1* showed compromised naive chemotaxis but adaptation remained intact (Bernhard and van der Kooy, 2000). These findings suggest that the underlying mechanism of adaptation to a chemosensory stimulus can be dissociated from the molecular process responsible for chemosensation itself.

B. Habituation

Although adaptation is a form of neural plasticity, it has been argued that it is not learning, but instead a type of receptor/cell fatigue. Learning theorists distinguish between non-associative habituation and fatigue or adaptation by how quickly the response shows recovery. In adaptation or fatigue, the only way to get the response back is to allow sufficient time for the response to recover. In contrast, habituation can be reversed immediately by the presence of a novel or noxious stimulus (Groves and Thompson, 1970; Rankin *et al.*, 1990). This induced rapid recovery is called dishabituation. Non-associative habituation to a chemosensory stimulus was first observed by Wen *et al.* (1997) who used prolonged pre-exposure to a relatively weak concentration of solubilized Na^+ as the habituating chemosensory stimulus. A smaller percentage of habituated worms migrated towards the Na^+ spot during the test compared to naive controls. Importantly, this decrease in approach behavior could be immediately reversed or dishabituated by a short exposure to a high concentration of Na^+ solution indicating that the change in approach behavior was habituation and not adaptation.

Bernhard and van der Kooy (2000) found that, in the case of diacetyl (an AWA neuron dependent chemical), whether response decrement is attributable to habituation or adaptation is determined by the concentration of the chemosensory stimulus. Pre-exposure to various concentrations of diacetyl followed by a chemotaxis assay showed that irreversible sensory adaptation occurred after high concentration pre-exposure, and that habituation occurred after low concentration pre-exposure. The decrement due to habituation could be reversed rapidly by a dishabituating stimulus (centrifugation twice for a total of 3 min at 250*g*), while this dishabituating stimulus had no effect on decreased responding due to adaptation. Thus, in the chemotaxis assay, habituation and adaptation to diacetyl can be dissociated by the concentration of the chemosensory stimulus.

C. State-Dependent Learning

C. elegans also possess the ability to make associations between a physiologic state and stimuli in their environment. In a study investigating the effects of starvation on olfactory adaptation, Colbert and Bargmann (1997) found that worms that were starved prior to pre-exposure to benzaldehyde showed greater sensory adaptation than worms that had not been starved. It was concluded that worms were learning an association between starvation and benzaldehyde exposure resulting in starved worms not migrating towards benzaldehyde for a longer period of time compared to unstarved controls. Surprisingly, the *adp-1* mutant strain (adaptation to benzaldehyde deficient), showed adaptation to benzaldehyde when starvation was paired with benzaldehyde pre-exposure. Further, the application of exogenous serotonin, which is known to be important for other feeding state-dependent behaviors (Avery and Horvitz, 1990; Croll, 1975; Horvitz *et al.*, 1982), blocks this starvation enhanced adaptation suggesting that the mechanism underlying benzaldehyde adaptation is somehow potentiated by the serotonin-dependent satiation pathway.

More recently, Nuttley *et al.* (2002) proposed that in the case of starvation-enhanced adaptation, starvation was acting as an aversive stimulus. If this were true, excessive amounts of food (*E. coli*) might act as an attractive stimulus, reversing any adaptation to benzaldehyde. When this hypothesis was tested, Nuttley *et al.* (2002) found that pre-exposure to benzaldehyde in the presence of three and six times the usual concentration of *E. coli* actually decreased adaptation to benzaldehyde. Further, a relatively high concentration of serotonin, which was used to mimic a greater food consumption, was sufficient to inhibit benzaldehyde adaptation compared to a lower concentration as used by Colbert and Bargmann (1997), which was used to mimic a normal feeding state. Because serotonin is thought to mimic a well-fed state, mutant strains were assayed that were known to have defective serotonin metabolism, such as *tph-1* (lacks tryptophan hydroxylase enzyme important for serotonin synthesis; Sze *et al.*, 2000) and *cat-4* (suggested to encode GTP cyclohydrolase I required for dopamine and serotonin synthesis; Nass and Blakely, 2003). Interestingly, neither of these mutants showed food-dependent inhibition of benzaldehyde adaptation, presumably because they could not respond to the presence of *E. coli*. When the administration of exogenous serotonin was paired with benzaldehyde, *cat-4* mutants showed decreased adaptation in the same way that wild-type worms showed decreased adaptation when benzaldehyde was paired with excessive food. Taken together, this evidence suggests a high complexity of neural plasticity in *C. elegans*; a simple behavior, such as benzaldehyde adaptation, can be state-dependent (associative) and this feeding state-dependent association is mediated by serotonin neurotransmission.

In addition to food, other substances can alter the internal state of an organism. For instance, ethanol produces an inebriated internal state; worms have been shown to be able to associate this ethanol-induced state with the adaptation to an odorant (Bettinger and McIntire, 2004). Similar to the aforementioned experiments, Bettinger and McIntire (2004) examined adaptation to benzaldehyde and found that pre-exposure to benzaldehyde coupled with intoxicating levels of ethanol resulted in inhibition of adaptation to benzaldehyde, if, and only if, testing occurred in the presence of intoxicating levels of ethanol; testing without ethanol or with low levels of ethanol caused normal benzaldehyde adaptation. Further, pre-exposure to benzaldehyde alone produced no decrease in adaptation when tested in the presence of ethanol, while ethanol pre-exposure had no effect on normal benzaldehyde chemotaxis. Bettinger and McIntire (2004) interpreted these findings to mean that worms were able to associate the internal ethanol-induced state with the presence of benzaldehyde and as such display increased adaptation to benzaldehyde only in the presence of ethanol. As well, ethanol-associative adaptation was shown to be odorant specific as ethanol-associated adaptation to benzaldehyde had no effect on butanone adaptation. Similar to Nuttley *et al.* (2002), Bettinger and McIntire (2004) found that mutant strains deficient in catecholamine transmission (*cat-1* and *cat-2*) were unable to demonstrate the ethanol-benzaldehyde association.

D. Classical Conditioning

Another form of associative learning termed differential classical conditioning has been demonstrated with chemotaxis in *C. elegans*. Wen *et al.* (1997) developed an assay similar to the chemotaxis assay previously described, but instead of a single attractive chemical and a control, the test plate contained two water-soluble attractants, Na^+ and Cl^- ions. Wen *et al.* (1997) produced classical conditioning by creating a preference for one of the ions (CS^+) by pairing it with another attractive stimulus, *E. coli* (US), followed by no food paired with the other ion (CS−). At test, worms were placed between the two ion point sources and conditioned worms significantly preferred the CS^+ ion that was paired with food (Wen *et al.*, 1997; for a summary of the procedure, see Fig. 3). These results were further supported by a second experiment that paired a repulsive US, garlic, with one of the ions. This time the worms showed a preference for the CS^- ion that was not paired with garlic (Wen *et al.*, 1997). A genetic screen identified two, as yet unidentified, mutants (*lrn-1* and *lrn-2*) that showed wild-type chemotaxis to both Na^+ and Cl^- ions and normal chemotactic habituation but were deficient in these associative learning tests (Wen *et al.*, 1997).

Saeki *et al.* (2001) described a modification of this procedure and showed that when worms were starved for 4 h on NaCl, their chemotaxis towards NaCl decreased. When the worms were starved in the presence of NaCl there was a decrease in chemotaxis towards other water-soluble chemoattractants sensed by ASE (Fig. 1). Application of serotonin partially blocked this conditioning suggesting that serotonin plays a role in the learning. Using this assay, three as yet unidentified mutant strains were isolated that showed lower levels of conditioning than wild-type worms.

The aforementioned classical conditioning results with taste have been extended using attractive odorants. Morrison *et al.* (1999) used diacetyl as the CS and acetic acid as an aversive US. When diacetyl was paired with acetic acid, worms showed a decreased chemotactic attraction to diacetyl. Similar to Wen *et al.* (1997), the *lrn-1* and *lrn-2* mutant strains could not learn this association, despite showing no deficits in naive chemotaxis, habituation, or dishabituation to diacetyl; however, spontaneous recovery after habituation was not fully analyzed. The results from these studies lead to the conclusion that *C. elegans* is capable of complex forms of chemosensory learning, and the mechanism by which the learning is processed is distinct from more simple non-associative learning, such as habituation.

From the research on chemosensory plasticity summarized thus far it has been shown that G-protein-coupled receptors (ODR-10; Bernhard and van der Kooy, 2000), TRP-like channels (OSM-9; Colbert *et al.*, 1997), and intracellular calcium signaling (Colbert and Bargmann, 1995) are implicated in simple chemosensory neuroplasticity, and catecholamines have been implicated in chemosensory learning during state-dependent paradigms (Bettinger and McIntire, 2004; Colbert and Bargmann, 1997; Nuttley *et al.*, 2002); however, as discussed in the mechanosensory learning section, glutamatergic neurotransmission has also been implicated in learning in *C. elegans* (Rankin and Wicks, 2000; Rose *et al.*, 2003). Kainate/AMPA-type glutamate receptors have been shown to play a role in learning during the performance of olfactory-mediated behaviors in other organisms (Larson *et al.*, 1995; Staubli *et al.*, 1996). The Kainate/AMPA-type receptor homolog in *C. elegans* is GLR-1, whose expression has been localized to the command interneurons for forward and backward movement (Hart *et al.*, 1995; Maricq *et al.*, 1995). Morrison and van der Kooy (2001) investigated the performance of *glr-1* mutants in both the non-associative and associative olfactory learning paradigms in order to evaluate the role of AMPA-like glutamate receptors. Using the volatile odorant chemotaxis assay, *glr-1* mutants were attracted to diacetyl in the same way as wild-type. Unlike wild-type, however, *glr-1* mutants did not show decreased chemotaxis towards diacetyl following conditioning where diacetyl was paired with the aversive acetic acid, suggesting that *glr-1* mutants are not capable of olfactory associative learning. This is similar to the *lrn-1* and *lrn-2* phenotype that Morrison *et al.* (1999) had observed. Unlike the *lrn-1* and *lrn-2* mutants, however, habituation to diacetyl was also deficient in the *glr-1* mutants. These findings suggests glutamate neurotransmission is important for both non-associative and associative forms of chemosensory learning and implicates the synapses between ASH and its targets as a potential site of plasticity important for mediating these behaviors.

Obviously, the integration of information from two different stimuli is important in mediating these associative forms of learning. Recently, a mutant found to be defective in proper sensory integration, called *hen-1*, also shows deficits in two learning paradigms (Ishihara *et al.*, 2002). Ishihara *et al.* (2002) developed an assay that measures the sensory integration of two chemosensory signals, diacetyl, which is an attractant, and Cu^{2+} ions, which is an aversive signal. By observing the behavior of a group of wild-type worms and their willingness to cross an aversive Cu^{2+} barrier to reach a source of diacetyl, measured by percentage of worms that cross the barrier, a characteristic dose-response curve was created for wild-type worms at varying concentrations of both stimuli. *hen-1* mutants showed abnormal curves, as they hesitated to cross the barrier unless diacetyl concentrations were very high, and Cu^{2+} concentrations were low, despite the fact that they showed normal chemotaxis towards/away from the two stimuli. Ishihara *et al.* (2002) argue that this abnormality in performance of this assay represents sensory integration deficits. Interestingly, these mutants also showed deficits in associative learning using NaCl as a CS and *E. coli* as a US, but showed regular wild-type behavior in NaCl chemotaxis. A second experiment confirmed this learning deficit because *hen-1* mutants also showed abnormal thermotaxis, which will be discussed further in the thermosensory learning and memory section. Ishihara *et al.* (2002) found that *hen-1* encodes for a secretory protein containing an LDL receptor motif that is expressed in ASE and AIY neurons (Fig. 1). These findings suggest that this secretory protein is important for the sensory integration necessary for associative forms of learning.

Finally, the most complex form of chemosensory learning demonstrated in *C. elegans* thus far, which involves a type of context conditioning known as occasion-setting, has recently been investigated using the olfactory adaptation paradigm (Law *et al.*, 2004). Law *et al.* (2004) used two types of chemosensory stimuli: benzaldehyde, an olfactory cue, and Na^+ or Cl^- ions, a taste cue, and found that the taste cue was salient enough to form contextual associations with the olfactory memory for the odor. They found that benzaldehyde adaptation was not as pronounced if worms were trained and tested using different taste cues compared to when both training and testing occurred using the same taste cues. They found that *che-1* mutants that do not properly develop the ASE neuron that is responsible for Na^+ and Cl^- reception, did not perform chemotaxis towards or adapt to these taste cues. Switching the taste context during benzaldehyde adaptation had no effect on the benzaldehyde chemotaxis index during the test of these mutants, suggesting that perception of the taste cues is essential for occasion-setting to occur.

Law *et al.* (2004) genetically dissociated associative adaptation and occasion-setting using the *gcp-1* and *lrn-2* mutants. GCP-1 is the gamma subunit of a G-protein-coupled receptor. *gcp-1* mutants show wild-type chemotaxis towards Na^+ and Cl^-, but exhibit adaptation deficits towards these two taste stimuli; however, *gcp-1* worms show an increased chemotaxis index for benzaldehyde when the contextual taste was switched from training to test during benzaldehyde adaptation compared to when the contextual taste was the same. This is consistent with the wild-type behavior, suggesting that despite the fact that *gcp-1* mutants cannot associate Na^+ and Cl^- with starvation, they can associate these taste cues with the memory of benzaldehyde adaptation. In contrast, *lrn-2* worms showed wild-type chemotaxis and adaptation to contextual stimuli, but performed equally well when the trained and tested context was the same compared to when it was different. This suggests that *lrn-2* mutants can form associations between the context stimulus and starvation, but cannot associate this occasion with the benzaldehyde adaptation. This is an odd finding considering that *lrn-2* was originally isolated in a screen for worms that cannot form associations between a taste and food (Wen *et al.*, 1997); *lrn-2* mutants were also deficient in forming associations between diacetyl and acetic acid (Morrison *et al.*, 1999). Now, in this assay *lrn-2* can form associations between a taste and benzaldehyde adaptation. From these experiments is not clear just what the specific nature of the learning deficits are in *lrn-2*.

E. Memory for Chemosensory Learning

So far only the acquisition (learning) of these chemosensory-based tasks has been discussed; however, there is a memory component as well. One critical feature about memory formation, previously discussed in the mechanosensory memory section, is the idea that in order for memory to be stored in a permanent way, protein synthesis must occur (Davis and Squire, 1984; Flood *et al.*, 1977; Frey *et al.*, 1988; Ghirardi *et al.*, 1995; Mirumori *et al.*, 1985; Montarolo *et al.*, 1986; Tully *et al.*, 1994). Morrison and van der Kooy (1997) investigated this idea using their odorant-based associative chemosensory conditioning paradigm by using cold shock, which was previously shown to block protein synthesis in *Drosophila* (Tully and Quinn, 1985), at different times following conditioning and testing retention 3 h later. Cold shock consisted of worms being cooled to $-20°$ C for 15 min either immediately, 15 min, or 1 h following conditioning. Conditioning involved the pairing of Na^+ or Cl^- ions (CS) with *E. coli* (US), and testing occurred either immediately or 3 h after conditioning. Cold-shock blocked memory when administered up to 30 min after conditioning when tested at 3 h. Morrison and van der Kooy (1997) suggested that administering cold shock after conditioning blocks memory retention for longer than 3 h, confirming the finding in most other organisms that longer lasting memory relies on protein synthesis.

Most recently, a genetic dissociation has been observed between acquisition (learning) and retrieval (memory) of chemosensory learning (Atkinson-Leadbeater *et al.*, 2004). Colbert and Bargmann (1995) showed that benzaldehyde pre-exposure causes both benzaldehyde and isoamyl alcohol adaptation, because the stimuli are generalized by the sensory neuron. A natural strain from Hawaii, CB4856 (different from the N2 strain usually used as wild-type) does not show benzaldehyde adaptation, but does show isoamyl alcohol adaptation following benzaldehyde pre-exposure. This finding suggests that the Hawaiian strain has the ability to generalize from benzaldehyde pre-exposure tested in this experiment by the isoamyl alcohol chemotaxis assay. However, CD4856 cannot retrieve the relevant information during the benzaldehyde chemotaxis test. Atkinson-Leadbeater *et al.* (2004) suggest this is because the Hawaiian *C. elegans* strain is capable of normal learning but has an olfactory memory retrieval deficit specific to benzaldehyde. This provides a potential model for studying learning and memory separately.

F. Summary of Chemosensory Learning and Memory

To summarize chemosensory learning and memory, simple forms of neural plasticity (Colbert and Bargmann, 1995), non-associative and associative learning (Wen *et al.*, 1997), and occasion-setting (Law *et al.*, 2004) have been observed in *C. elegans* in response to chemical stimuli. These changes in behavior can be modified by state dependency, such as feeding and ethanol-induced states (Bettinger and McIntire, 2004; Colbert and Bargmann, 1997; Nuttley *et al.*, 2002). The clues that hint at the molecular mechanisms that mediate chemosensory learning include the action of G-protein-coupled receptors (Bernhard and van der Kooy, 2000; Law *et al.*, 2004), TRP-like channels (Colbert *et al.*, 1997), an LDL receptor-like secretory protein (Ishihara *et al.*, 2002), intracellular calcium signaling (Colbert and Bargmann, 1995), glutamate (Morrison and van der Kooy, 2001) and catecholamine (Bettinger and McIntire, 2004; Colbert and Bargmann, 1997; Nuttley *et al.*, 2002) neurotransmission, and dependency on protein synthesis (Morrison and van der Kooy, 1997). More research needs to be completed to find the link between all of these clues to form a reasonable mechanism of action.

IV. Thermosensory Learning and Memory

Thermotaxis behavior is characterized as movement towards or away from a thermal stimulus. This behavior was first characterized by Hedgecock and Russell (1975) who found that worms cultivated in a well-fed environment at a constant temperature migrate to their cultivation temperature if transferred to a thermal gradient. If placed on a radial thermal gradient, worms will show isothermal tracking within 3°C at the cultivation temperature. Conversely, if nutritionally deprived (by food depletion, overcrowding, and so on) worms will avoid the cultivation temperature when placed in a thermal gradient (Hedgecock and Russell, 1975). A mutant screen was developed to identify abnormal thermotaxis. Mutant strains were grouped into three different phenotypic categories: thermophilic, which migrate towards temperatures that are greater than their cultivation temperature; cryophilic, which migrate towards temperatures that are less than their cultivation temperature; and athermotactic, which move through the thermal gradient randomly, not showing a preference for any area of temperature (Hedgecock and Russell, 1975; for a summary of phenotypes, see Fig. 4). Over the years, a plethora of mutants with thermotactic deficits have been characterized including *tax-2*, *tax-4*, *ttx-3*, and *lin-11*, which have mutations in genes that encode cyclic nucleotide-gated channels and transcription factors (for a review, see Mori, 1999).

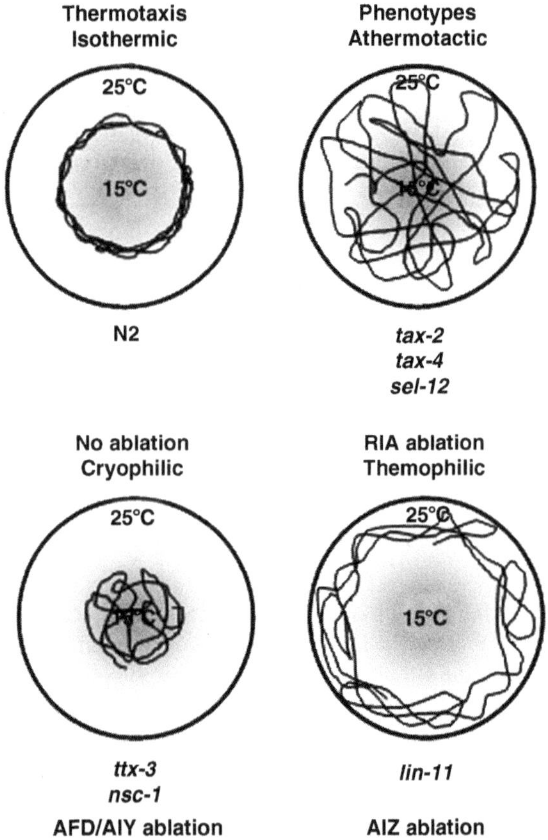

FIG. 4. Paths of mutant strains and laser ablated worms on an agar filled petri dish with a radial thermal gradient of 15°C to 25°C after being cultivated in a well-fed environment at 20°C. Phenotypes are describes as: isothermal when worms track along the cultivation temperature (intact wild-type worms; N2); athermotactic when worms do not show a preferred temperature (*tax-2*, *tax-4*, *sel-12* and RIA ablations); cryophilic when worms migrate to temperatures cooler than the cultivation temperature (*ttx-3*, *ncs-1* and AFD/AIY ablations); and thermophilic when worms migrate to temperatures warmer than the cultivation temperature (*lin-11* and AIZ ablations).

A. Neural Circuit Analysis

In 1995, the neural circuit for this behavior was mapped using laser ablation (Mori and Oshima, 1995) (Fig. 1). In order to determine the effects on wild-type thermotactic behavior, the destruction of individual neurons was necessary. This approach identified the pair of AFD sensory neurons, which have finger-like ciliated sensory endings enclosed by the amphid sheath cell. AFD projects to an interneuron, AIY. In addition to AIY, another interneuron, AIZ, has also been identified to be important for normal thermotactic behavior. AIY-ablated worms exhibit a cryophilic phenotype, while worms with ablated AIZ neurons respond thermophilically. This dissociation of thermotactic function suggests that activation of these interneurons results in behavioral drives that are balanced in wild-type such that when the function of one neuron is eliminated through ablation the other pathway determines temperature-directed behavior. Ablation of RIA, an interneuron located in the nerve ring of *C. elegans*, caused athermotactic behavior and as such has been considered a site of thermophilic and cryophilic integration (Mori and Oshima, 1995). Other neurons of minor importance were also identified; for example, AFD has been shown to also project to AIB, an interneuron also thought to support the cryophilic drive. Mori and Oshima (1995) constructed a neural circuit diagram using this data (simplified and integrated into Fig. 1).

B. Genes Involved in Thermotactic Learning

Mori (1999) suggested that this thermotactic behavior would be a good model for studying learning and memory because, in order to perform proper thermotaxis, worms must first encode a cultivation temperature into memory and then pair this information with the appropriate feeding state (feeding or starvation), a form of associative learning. This information must then be recalled upon, transferred to a thermal gradient, and compared to the current temperature, at which point an appropriate response can be made. Recent studies are now finding clues that will help to elucidate the cellular and molecular mechanisms that mediate this behavior.

One protein that has been found to be expressed in the AIY neuron is neuronal calcium sensor-1 (NCS-1) (Gomez *et al.*, 2001). This protein is also expressed in AFD, and eight other pairs of amphid sensory neurons important for taste and olfactory sensation; however, *ncs-1* mutants do not show abnormal taste and olfactory-mediated behaviors (Gomez *et al.*, 2001). When tested using the same thermotactic learning assay previously described, *ncs-1* worms were observed to mainly perform cryophilically (Gomez *et al.*, 2001). This is congruent with AFD and AIY-ablated worms as well as mutations in TTX-3, which codes for a transcription factor that is specific to AIY neurons (Mori, 1999). An AIY-specific promoter rescue of *ncs-1* led to a full behavioral recovery, while an AFD-specific promoter rescue of *ncs-1* did not; therefore, the site of action of NCS-1 was narrowed to the AIY neuron (Gomez *et al.*, 2001). Because NCS-1 is a calcium-sensing protein, Gomez *et al.* (2001) asked whether the NCS-1–mediated thermotactic learning was calcium-binding–dependent. This was addressed by expressing an *ncs-1* transgene with a single-point mutation that made the calcium-binding region non-functional, in an *ncs-1* knockout background. Worms with this loss-of-function rescue did not recover wild-type thermotactic behavior (Gomez *et al.*, 2001). This data suggests that intracellular calcium signaling is an important element in mediating thermotactic learning through the AIY neuron.

C. Memory for Cultivation Temperature

As memory for a temperature associated with food can be maintained for hours, it suggests that the memory results in a consistent neuronal change. Samuel *et al.* (2003) used fluorescence recovery after photobleaching (FRAP) of a pH-sensitive GFP localized to synaptic vesicles to investigate whether neurotransmitter release from AFD changes during thermotaxis. Samuel *et al.* (2003) found greater fluorescence recovery ($>50\%$) in worms that experienced temperatures different (either higher or lower) from their respective cultivation temperature. When the current temperature was the same as the worm's cultivation temperature, this resulted in lower ($<10\%$) fluorescence recovery suggesting the AFD neuron was not activated in this condition. In another study, Kimura *et al.* (2004) measured calcium transients in AFD by imaging cameleon, a genetically encodable calcium sensor comprised of a chimeric protein containing cyan fluorescent protein (CFP) and yellow fluorescent protein (YFP) derivatives and a calcium-binding domain. When calcium binds to the calcium-binding domain, this results in fluorescence resonance energy transfer (FRET) between CFP and YFP, and a concomitant change in fluorescence ratio between CFP and YFP indicative of calcium concentration. Kimura *et al.* (2004) focused on calcium transients in the AFD cell body and found increases in calcium concentration only when worms were experiencing temperatures higher than their cultivation temperature. Although these results do not wholly agree with what Samuel *et al.* (2003) reported, both papers do provide evidence for a change in AFD properties during thermotaxis. These changes could represent an aspect of the molecular substrates of thermotactic learning and memory.

Thermotaxis has also been used to characterize homologues of human presenilin genes, which play a role in human familial Alzheimer's disease. A strain with a mutation in one of these presenilin homologues, *sel-12*, has been found to have abnormal AIY neuron morphology and shows abnormal thermotactic memory; *sel-12* worms are mostly athermotactic with 10% of mutants showing cryophilic behavior (Wittenburg *et al.*, 2000) suggesting they do not remember the cultivation temperature. *sel-12* and PS1 (human presenilin) transgenes rescued the thermotactic behavior as well as the defects in AIY morphology seen with the *sel-12* mutation. Understanding how the molecular mechanisms of thermotaxis are compromised in *sel-12* worms may provide clues with regards to mechanisms of the progression of Alzheimer's disease.

Once a memory for a food-paired temperature is established, it is not resistant to further modification. In their original report on thermotaxis, Hedgecock and Russell (1975) demonstrated an element of plasticity to thermotaxis in that the preferred temperature could be reset to a temperature that is different to the cultivation temperature. To investigate this, Hedgecock and Russell (1975) cultivated worms with food at one temperature and then temperature-shifted as adults to a new temperature. In this case, worms raised at 16° C and then shifted to 25° C as adults showed a preference for the new temperature within 2 to 4 h by tracking the isotherm of the cultivation temperature. The same was found to be true for worms that were shifted from a higher cultivation temperature to a new lower temperature (Hedgecock and Russell, 1975). This process of associating a new temperature with food appeared to be time-dependent rather than temperature-specific as both up- and down-shifted worms showed behavioral changes within approximately the same time frame.

D. Dissecting Thermotactic Learning

Recently, Mohri *et al.* (2005) further characterized the thermotactic learning process by separating the isothermal tracking behavior into its components: response to a temperature shift and change in behavior due to presence of food. Mohri *et al.* (2005) accomplished this by examining the behavior of worms that were switched from their cultivation temperature into a different temperature that also contained food to observe how long it took to learn the new temperature. In a second experiment, the temperature remained constant, but the worms were transferred to a different feeding state, to determine how long it took to learn a new feeding state. They found that it took 2 to 3 h to learn the new temperature, and this learning was temperature-independent. The worms learned the new feeding state in a temperature-dependent fashion, when transferred from a well-fed to a starved environment. At a cool temperature (17°C), they learned to avoid the temperature after 3 h, while for worms cultivated at a warmer temperature (25° C), it only took 30 min to learn. In the opposite switch, when worms were transferred from starved to well-fed conditions, worms learned to migrate to the appropriate isotherm in <10 min. These results suggest that encoding the environmental temperature is a relatively long process (2 to 4 h), while learning to pair the feeding state to the stored temperature is a relatively quick process (10 to 30 min). A final experiment involved a double switch of temperature and feeding state. Worms initially associated the new feeding state with the old temperature and then gradually learned the new temperature over a similar time scale as the previous experiments. For example, worms that were starved at 25° C initially avoided 25° C on the thermal gradient test plate, but only 10 min after being switched to a bacteria-seeded 17° C cultivation plate, the worms migrated towards the 25° C isotherm during the test. After 1 h, however, they had learned the new association.

To further dissect the two components of the behaviors, Mohri *et al.* (2005) examined the role that different neurotransmitters play in thermotaxis. Serotonin and octopamine are two neurotransmitters that have been associated with other feeding state-dependent behaviors in *C. elegans*, such as egg laying and locomotion (Avery and Horvitz, 1990; Croll, 1975; Horvitz *et al.*, 1982). Exogenous application of serotonin can be thought to mimic a well-fed state; exogenous treatment of serotonin to starved worms during conditioning leads to attraction to the cultivation temperature. Conversely, octopamine can be thought to mimic starving conditions; well-fed worms treated with exogenous octopamine during cultivation avoid the cultivation temperature. This suggests a role of these two neurotransmitters in the associative learning necessary for thermotaxis.

Mohri *et al.* (2005) also performed a genetic screen in order to find strains that show abnormal thermotactic learning as expressed by isothermal tracking, using this new dissection of the behavior. Three mutants were identified to perform normal thermotaxis after being cultivated in a well-fed state, but would not avoid the cultivation temperature if conditioned in a starved state. This suggests that the mutants could encode the cultivation temperature, but were deficient at associating this encoded temperature with the conditioned feeding state. They were designated *a*bnormal *h*unger *o*rientation, *aho-1*, *aho-2*, and *aho-3*. *aho-1* and *aho-3* mutants only showed this abnormal hunger phenotype at 17°C and 25°C, respectively, and, therefore, were temperature-dependent. *aho-2* was temperature-independent. Other sensory behaviors were investigated in these mutants in order to elucidate the site and mechanism of this thermotactic associative learning. *aho-1* was slightly defective at NaCl chemotaxis, and *aho-3* was deficient at male mating, but all other sensory phenotypes that were tested were wild-type. This study by Mohri *et al.* (2005) demonstrates that there are two aspects to thermotaxis, thermal memory and associative learning, and that these are distinct and dissociable both behaviorally and genetically.

As discussed in the previous section, *hen-1* mutants, which showed both a chemosensory integration and learning deficits, also show deficits in isothermal tracking (Ishihara *et al.*, 2002). Interestingly, the *hen-1* phenotype is the same as the *aho* mutants: *hen-1* mutants will migrate to the cultivation temperature regardless of the feeding state that was previously paired with the cultivation temperature. This provides further support for the role of HEN-1, an LDL receptor-like secretory protein, in sensory integration-mediated learning, and more importantly changes its role from being strictly necessary for chemosensory integration to being involved in a more general sensory integration mechanism (Ishihara *et al.*, 2002).

Recent evidence suggests that there is an age-related decline in isothermal tracking (Murakami and Murakami, 2005). In addition, worms with mutations in genes like *clk-1* and *isp-1*, which play a role in decreasing oxidative stress, showed less age-related decline than wild-type worms. Worms with mutations that led to increased oxidative stress (*mev-1* and *gas-1*) showed severely impaired learning. Thus, it appears as though oxidative stress may be an important factor in a worm's ability to show appropriate isothermal tracking.

E. Summary of Thermotactic Learning

To summarize thermosensory learning and memory, Hedgecock and Russell (1975) were the first to characterize thermotaxis and observed that the behavior was plastic. The neural circuit was identified using laser ablation techniques, at which point a genetic screen was developed isolating numerous mutant strains (Mori, 1999; Mori and Oshima, 1995). Further characterization of the plasticity of isothermal tracking has just recently been conducted concluding with the identification of more mutants showing that the temperature encoding and the associative pairing necessary for thermotaxis can be genetically dissociated (Mohri *et al.*, 2005). Clues to the cellular and molecular mechanisms mediating thermotaxis include the localization of the potential site of plasticity at the AFD and AIY neurons (Gomez *et al.*, 2001; Kimura *et al.*, 2004; Samuel *et al.*, 2003), and the role of certain molecules, such as NCS-1 (Gomez *et al.*, 2001) and HEN-1 (Ishihara *et al.*, 2002).

V. Conclusions

Learning and memory in *C. elegans* was first described using a simple non-associative mechanosensory training paradigm, habituation of the tap withdrawal response (Rankin *et al.*, 1990). Since then learning has been shown in multiple sensory modalities including chemosensation and thermosensation. More complex forms of learning and memory have been investigated, and it is now obvious that *C. elegans* can learn classical conditioning, context conditioning, and occasion-setting. The molecular mechanisms for these processes are being investigated with many clues already apparent. Some of this evidence includes the localization of the sites of plasticity, the neurotransmitters involved, and the molecules important for intracellular signaling within the neurons. With the recent advances in imaging technology, *C. elegans* has become one of the best simple animal models for studying the cellular and molecular mechanisms that mediate learning and memory. Because of the vast similarities between the elements of *C. elegans* nervous system and the nervous system of other, more complex organisms, experimental findings in *C. elegans* could lead to important clues concerning learning and memory in general.

References

Atkinson-Leadbeater, K., Nuttley, N. M., and van der Kooy, D. (2004). A genetic dissociation of learning and recall in *Caenorhabditis elegans*. *Behav. Neurosci.* **18,** 1206–1213.

Avery, L., and Horvitz, H. R. (1990). Effects of starvation and neuroactive drugs on feeding in *Caenorhabditis elegans*. *J. Exp. Zoology* **253,** 263–270.

Bargmann, C. I., Hartwieg, E., and Horvitz, H. R. (1993). Odorant-selective genes and neurons mediate olfaction in *C. elegans*. *Cell* **74,** 515–527.

Bargmann, C. I., and Mori, I. (1997). Chemotaxis and thermotaxis. *In* "C. elegans II" (D. L. Riddle, T. Blumenthal, B. J. Meyer, and J. R. Preiss, Eds.), pp. 717–737. Cold Spring Harbor Laboratory Press, New York.

Beck, C. D. O., and Rankin, C. H. (1993). Effects of aging on habituation in the nematode *Caenorhabditis elegans*. *Behav. Proc.* **28,** 145–164.

Beck, C. D. O., and Rankin, C. H. (1995). Heat shock disrupts long-term memory consolidation in *Caenorhabditis elegans*. *Learning Memory* **2,** 161–177.

Beck, C. D. O., and Rankin, C. H. (1997). Long-term habituation is produced by distributed training at long ISIs and not by massed training at short ISIs in *Caenorhabditis elegans*. *Animal Learning Behavior* **25,** 446–457.

Bellocchio, E. E., Reimer, R. J., Fremeau, R. T., Jr., and Edwards, R. H. (2000). Uptake of glutamate into synaptic vesicles by an inorganic phosphate transporter. *Science* **289,** 957–960.

Bernhard, N., and van der Kooy, D. (2000). A behavioural and genetic dissection of two forms of olfactory plasticity in *Caenorhabditis elegans*: Adaptation and habituation. *Learning Memory* **7,** 199–212.

Bettinger, J. C., and McIntire, S. L. (2004). State dependency in *C. elegans*. *Gene Brain Behav.* **3,** 266–272.

Broster, B. S., and Rankin, C. H. (1994). Effects of changing interstimulus interval during habituation in *Caenorhabditis elegans*. *Behav. Neurosci.* **108,** 1019–1029.

Brockie, P. J., Madsen, D. M., Zheng, Y., Mellem, J., and Maricq, A. V. (2001). Differential Expression of glutamate receptor subunits in the nervous system of *Caenorhabditis elegans* and their regulation by the homeodomain protein UNC-42. *J. Neurosci.* **21,** 1510–1522.

Carew, T. J., Pinsker, H. M., and Kandel, E. R. (1972). Long-term habituation of a defensive withdrawal reflex in *Aplysia*. *Science* **175,** 451–454.

Chalfie, M., Sulston, J. E., White, J. G., Southgate, E., Thompson, J. N., and Brenner, S. (1985). The neural circuit for touch sensitivity in *Caenorhabditis elegans*. *J. Neurosci.* **5,** 956–964.

Chiba, C. M., and Rankin, C. H. (1990). An analysis of the development of the reversal reflex in *Caenorhabditis elegans*. *J. Neurobiol.* **21,** 543–554.

Colbert, H. A., and Bargmann, C. I. (1995). Odorant-specific adaptation pathways generate olfactory plasticity in *C. elegans*. *Neuron* **14,** 803–812.

Colbert, H. A., and Bargmann, C. I. (1997). Environmental signals modulate olfactory acuity, discrimination, and memory in *Caenorhabditis elegans*. *Learning Memory* **4,** 179–191.

Colbert, H. A., Smith, T. L., and Bargmann, C. I. (1997). OSM-9, a novel protein with structural similarity to channels, is required for olfaction, mechanosensation, and olfactory adaptation in *Caenorhabditis elegans*. *J. Neurosci.* **17,** 8259–8269.

Croll, N. A. (1975). Indolealkylamines in the coordination of nematode behavioral activities. *Can J. Zoology* **53,** 894–903.

Davis, H. P., and Squire, L. R. (1984). Protein synthesis and memory: A review. *Psychol. Rev.* **96,** 518–559.

Ebbinghaus, H. (1885). Retention as a function of repeated learning. *In* Memory (reprint edition, 1964), pp. 81–89. Dover: New York.

Evans, J. G. M., and Hammond, G. R. (1983). Differential generalization of habituation across contexts as a function of stimulus significance. *Animal Learning Memory* **11,** 431–434.

Flood, J. F., Bennett, E. L., Orme, A. E., and Jarvik, M. E. (1977). Protein synthesis gradient of ECS retrograde amnesia. *Behav. Biol.* **21,** 307–328.

Frey, U., Krug, M., Reymann, K. G., and Matthies, H. (1988). Anisomycin, and inhibitor of protein synthesis, blocks late phases of LTP phenomena in the hippocampal CA1 region *in vitro*. *Brain Res.* **452,** 57–65.

Ghirardi, M., Montarolo, P. G., and Kandel, E. R. (1995). A novel intermediate stage in the transition between short- and long-term facilitation in the sensory to motor neuron synapse of *Aplysia*. *Neuron* **14,** 413–420.

Gomez, M., De Castro, E., Guarin, E., Sasakura, H., Kuhara, A., Mori, I., Bartfai, T., Bargmann, C. I., and Nef, P. (2001). Ca^{2+} Signaling via the neuronal calcium sensor-1 regulates associative learning and memory in *C. elegans*. *Neuron* **30,** 241–248.

Groves, P. M., and Thompson, R. F. (1970). Habituation: A dual-process theory. *Psychol. Rev.* **77,** 419–450.

Hart, A. C., Sims, S., and Kaplan, J. M. (1995). Synaptic code for sensory modalities revealed by *C. elegans* GLR-1 glutamate receptor. *Nature* **378,** 82–85.

Hedgecock, E. M., and Russell, R. L. (1975). Normal and mutant thermotaxis in the nematode *Caenorhabditis elegans*. *Proc. Natl. Acad. Sci.* **72,** 4061–4065.

Hobert, O. (2003). Behavioral plasticity in *C. elegans*: Paradigms, circuits, genes. *J. Neurobiol.* **54,** 203–223.

Horvitz, H. R., Chalfie, M., Trent, C., Sulston, J. E., and Evans, P. D. (1982). Serotonin and octopamine in the nematode *Caenorhabditis elegans*. *Science* **216,** 1012–1014.

Ishihara, T., Iino, Y., Mohri, A., Mori, I., Gengyo-Ando, K., Mitani, S., and Katsura, I. (2002). HEN-1, a secretory protein with an LDL receptor motif, regulates sensory integration and learning in *Caenorhabditis elegans*. *Cell* **109,** 639–649.

Kimura, K. D., Miyawaki, A., Matsumoto, K., and Mori, I. (2004). The *C. elegans* thermosensory neuron AFD responds to warming. *Curr. Biol.* **14,** 1291–1295.

Larson, J., Lieu, T., Petchpradub, V., LeDuc, B., Ngo, H., Rogers, G. A., and Lynch, G. (1995). Facilitation of olfactory learning by a modulator of AMPA receptors. *J. Neurosci.* **15,** 8023–8030.

Law, E., Nuttley, W. M., and van der Kooy, D. (2004). Contextual taste cues modulate olfactory learning in *C. elegans* by an occasion-setting mechanism. *Curr. Biol.* **14,** 1303–1308.

Lee, R. Y. N., Sawin, E. R., Chalfie, M., Horvitz, H. R., and Avery, L. (1999). EAT-4, a homolog of a mammalian sodium-dependent inorganic phosphate cotransporter, is necessary for glutamatergic neurotransmission in *Caenorhabditis elegans*. *J. Neurosci.* **19,** 159–167.

Lüscher, C., and Frerking, M. (2001). Restless AMPA receptors: Implications for synaptic transmission and plasticity. *Trends Neurosci.* **24,** 665–670.

Malinow, R., and Malenka, R. C. (2002). AMPA receptor trafficking and synaptic plasticity. *Ann. Rev. Neurosci.* **25,** 103–126.

Maricq, A. V., Peckol, E., Driscoll, M., and Bargmann, C. I. (1995). Mechanosensory signaling in *C. elegans* mediated by the GLR-1 glutamate receptor. *Nature* **378,** 78–81.

Mirumori, S. J. Y., Rosenzweig, M. R., and Bennett, E. L. (1985). Long-term working memory in the rat: Effects of hippocampally applied anisomycin. *Behav. Neurosci.* **99,** 220–232.

Mohri, A., Kodama, E., Kimura, K. D., Koike, M., Mizuno, T., and Mori, I. (2005). Genetic control of temperature preference in the nematode *Caenorhabditis elegans. Genetics* **169,** 1437–1450.

Montarolo, P. G., Goelet, P., Castellucci, V. F., Morgan, J., Kandel, E. R., and Schacher, S. (1986). A critical period for macromolecular synthesis in long-term heterosynaptic facilitation in *Aplasia. Science* **234,** 1249–1254.

Mori, I. (1999). Genetics of chemotaxis and thermotaxis in the nematode *Caenorhabditis elegans. Ann. Rev. Genetics* **33,** 399–422.

Mori, I., and Oshima, Y. (1995). Neural regulation of thermotaxis in Caenorhabditis elegans. *Nature* **376,** 344–348.

Morrison, G. E., Wen, J. Y., Runciman, S., and van der Kooy, D. (1999). Olfactory associative learning in *Caenorhabditis elegans* is impaired in *lrn-1* and *lrn-2* mutants. *Behav. Neurosci.* **113,** 358–367.

Morrison, G. E., and van der Kooy, D. (1997). Cold shock before associative conditioning blocks memory retrieval, but cold shock after conditioning blocks memory retention in *Caenorhabditis elegans. Behav. Neurosci.* **111,** 564–578.

Morrison, G. E., and van der Kooy, D. (2001). A mutation in the AMPA-type glutamate receptor, *glr-1,* blocks olfactory associative and nonassociative learning in *Caenorhabditis elegans. Behav. Neurosci.* **115,** 640–649.

Murakami, S., and Murakami, H. (2005). The effects of aging and oxidative stress on learning behavior in *C. elegans. Neurobiol. Aging* **26,** 899–905.

Nass, R., and Blakely, R. D. (2003). The *Caenorhabditis elegans* dopaminergic system: Opportunities for insights into dopamine transport and neurodegeneration. *Ann. Rev. Pharmacol. Toxicol.* **43,** 521–544.

Nuttley, W. M., Atkinson-Leadbeater, K. P., and van der Kooy, D. (2002). Serotonin mediates food-odor associative learning in the nematode *Caenorhabditis elegans. Proc. Natl. Acad. Sci. USA* **99,** 12449–12454.

Rankin, C. H. (2000). Context conditioning in habituation in the nematode *C. elegans. Behav. Neurosci.* **114,** 496–505.

Rankin, C. H., Beck, C. D. O., and Chiba, C. M. (1990). *Caenorhabditis elegans*: A new model system for the study of learning and memory. *Behav. Brain Res.* **37,** 89–92.

Rankin, C. H., and Broster, B. S. (1992). Factors affecting habituation and recovery from habituation in the nematode *Caenorhabditis elegans. Behav. Neurosci.* **106,** 239–242.

Rankin, C. H., Gannon, T., and Wicks, S. R. (2000). Developmental analysis of habituation in the nematode *C. elegans. Develop. Psychobiol.* **36,** 261–270.

Rankin, C. H., and Wicks, S. R. (2000). Mutations of the *C. elegans* brain-specific inorganic phosphate transporter, *eat-4,* affect habituation of the tap-withdrawal response without affecting the response itself. *J. Neurosci.* **20,** 4337–4344.

Rose, J. K., Kaun, K. R., and Rankin, C. H. (2002). A new group training procedure for habituation demonstrates that presynaptic glutamate release contributes to long-term memory in *C. elegans. Learning Memory* **9,** 130–137.

Rose, J. K., Kaun, K. R., Chen, S. H., and Rankin, C. H. (2003). GLR-1, a non-NMDA glutamate receptor homolog, is critical for long-term memory in *Caenorhabditis elegans. J. Neurosci.* **23,** 9595–9599.

Saeki, S., Yamamoto, M., and Ino, Y. (2001). Plasticity of chemotaxis revealed by paired presentation of a chemoattractant and starvation in the nematode *Caenorhabditis elegans*. *J. Exp. Biol.* **204,** 1757–1764.

Samuel, A. D. T., Silva, R. A., and Murthy, V. N. (2003). Synaptic activity of the AFD neuron in *Caenorhabditis elegans* correlates with thermotactic memory. *J. Neurosci.* **23,** 373–376.

Sanyal, S., Wintle, R. F., Kindt, K. S., Nuttley, W. M., Arvan, R., Fitzmaurice, P., Bigras, E., Merz, D. C., Hébert, T. E., van der Kooy, D., Schafer, W. R., Culotti, J. G., and Van Tol, H. H. M. (2004). Dopamine modulates the plasticity of mechanosensory responses in *Caenorhabditis elegans*. *EMBO J.* **23,** 473–482.

Sengupta, P., Chou, J. H., and Bargmann, C. I. (1996). *odr-10* encodes a seven transmembrane domain olfactory receptor required for responses to the odorant diacetyl. *Cell* **84,** 899–909.

Staubli, U., Izrael, Z., and Xu, F. (1996). Remembrance of odors past: Enhancement by central facilitation of AMPA receptors. *Behavl Neurosci.* **110,** 1067–1073.

Sze, J. Y., Victor, M., Loer, C., Shi, Y., and Ruvkun, G. (2000). Food and metabolic signaling defects in a *Caenorhabditis elegans* serotonin-synthesis mutant. *Nature* **403,** 560–564.

Tully, T., Preat, T., Boynton, S. C., and Del Vecchio, M. (1994). Genetic dissection of consolidated memory in *Drosophila*. *Cell* **79,** 35–47.

Tully, T., and Quinn, W. G. (1985). Classical conditioning and retention in normal and mutant *Drosophila melanogaster*. *J. Comp. Physiol. [A] Sensory Neural Behavl. Physiol.* **157,** 263–277.

Wagner, A. R. (1976). Priming in STM: An information-processing mechanism for self-generated or retrieval-generated depression in performance. *In* "Habituation: Perspective From Child Development, Animal Behavior, and Neurophysiology" (T. J. Tighe and R. N. Leaton, Eds.), pp. 95–128. Hillsdale, Erlbaum.

Wagner, A. R. (1978). Expectancies and priming of STM. *In* "Cognitive Processes in Animal Behavior" (S. H. Hulse, H. Fowler, and W. R. Honig, Eds.), pp. 177–209. Hillsdale, Erlbaum.

Wagner, A. R. (1979). Habituation and memory. *In* "Mechanisms of Learning and Motivation" (A. Dickinson and R. A. Boakes, Eds.), pp. 53–82. Hillsdale, Erlbaum.

Ward, S. (1973). Chemotaxis by the nematode *Caenorhabditis elegans*: Identification of attractants and analysis of the response by use of mutants. *Proc. Natl. Acad. Sci. USA* **70,** 817–821.

Wen, J. Y. M., Kumar, N., Morrison, G., Rambaldini, G., Runciman, S., Rousseau, J., and Van der Kooy, D. (1997). Mutations that prevent associative learning in *C. elegans*. *Behav. Neurosci.* **111,** 354–368.

Wicks, S. R., and Rankin, C. H. (1995). Integration of mechanosensory stimuli in *Caenorhabditis elegans*. *J. Neurosci.* **15,** 2434–2444.

Wicks, S. R., and Rankin, C. H. (1996). Recovery from habituation in *Caenorhabditis elegans* is dependent on interstimulus interval and not habituation kinetics. *Behav. Neurosci.* **4,** 840–844.

Wicks, S. R., and Rankin, C. H. (1997). The effects of tap withdrawal response habituation on other withdrawal behaviors: The localization of habituation in *C. elegans*. *Behav. Neurosci.* **111,** 1–12.

Wittenburg, N., Eimer, S., Lakowski, B., Röhrig, S., Rudolph, C., and Baumeister, R. (2000). Presenilin is required for proper morphology and function of neurons in *C. elegans*. *Nature* **406,** 306–309.

Xu, X., Sassa, T., Kunoh, K., and Hosono, R. (2002). A mutant exhibiting abnormal habituation behavior in *Caenorhabditis elegans*. *J. Neurogenet.* **16,** 29–44.

NEURAL SPECIFICATION AND DIFFERENTIATION

Eric Aamodt* and Stephanie Aamodt[†]

*Department of Biochemistry, Louisiana State University Health Sciences Center-Shreveport Shreveport, Louisiana

[†]Department of Biological Sciences, Louisiana State University in Shreveport Shreveport, Louisiana

Neurogenesis studies in simple, genetically tractable model organisms, such as *Caenorhabditis elegans* and *Drosophila*, are beginning to uncover principles common to neurogenesis in all animals. The genes involved in neurogenesis are often classified into proneural, neurogenic, pan-neural, and neuronal-type selector genes, but many appear to act at multiple levels. In *C. elegans*, a proneural gene, like *lin-32*, which encodes a bHLH transcription factor, regulates lineages that generate neurons, is proneural, and also carries neuronal sub-type information. A gene like *unc-86*, which encodes a POU-homeodomain transcription factor, controls lineages, is required for neurogenesis, activates the subtype selector gene *mec-3* and together with *mec-3* activates the genes that actually build the touch receptor neurons, such as the ion channel gene *mec-4* and the beta-tubulin gene *mec-7*. On the other hand, the LIM-homeodomain transcription factors are simpler. They are expressed after the last cell division and regulate neuronal

INTERNATIONAL REVIEW OF NEUROBIOLOGY, VOL. 69
DOI: 10.1016/S0074-7742(05)69003-4

0074-7742/06 $35.00

subtype specific genes. There are also genes that regulate a subset of the genes required in specific types of neurons. The logic of the transcriptional regulatory networks that contain these genes is beginning to emerge.

I. Introduction

Neurogenesis is necessarily complex due to the vast array of neuronal cell types. To function correctly, each neuron must be properly positioned, establish the proper axonal projections, have the proper dendritic morphology, and must express the proper ion channels, neurotransmitters, and neurotransmitter receptors. The consequences of this complex process are the remarkable behavioral and cognitive repertoires of animals, which certainly justify the effort required to understand it.

During the past century, genetic screens, done mostly in *Drosophila* and *Caenorhabditis elegans*, have identified numerous genes involved in neurogenesis. Table I lists many of the *C. elegans* genes known to be involved in neurogenesis, their loss of function phenotypes, homologs in *Drosophila* and/or mammals, and the type of gene product they produce. These genes have been classified based on the point in neurogenesis where they act. Proneural genes endow the cells that express them with the potential to become neurons. Neurogenic genes single out particular cells to become neurons. Pan-neural precursor genes are expressed in all neural precursors. Neuronal-type selector genes are expressed in subsets of neurons. As more has been learned about the roles of these genes in neurogenesis it has become clear that many of these genes act at multiple levels.

Analysis of the vertebrate homologues of these neurogenesis genes suggests that many of the mechanisms involved in neurogenesis are conserved. In this chapter, we will consider the regulation of neuronal development in *C. elegans* in light of what is known about neurogenesis in these other systems.

C. elegans is an excellent system for identifying mutations that affect the nervous system. There are powerful genetic tools available in this system including both forward and reverse genetic approaches, and mutations that damage much of the nervous system are viable. The animals can feed and survive without moving or sensing their environment.

There are 302 neurons in the *C. elegans* adult hermaphrodite that fall into 118 classes based on morphology, connectivity, and position (White *et al.*, 1986). All 302 neurons, their morphology, synapses, and gap junctions have been reconstructed from serial section electron micrographs (Albertson and Thomson, 1976; White *et al.*, 1976, 1986). Sensory and interneurons are mostly bilaterally symmetric pairs. Approximately one-half of the neurons are in the head surrounding a central neuropile called the nerve ring. The pharynx has its own nervous system containing 20 neurons and it is connected to the rest of the nervous system by a

TABLE I
C. elegans Neurogenesis Genes

Gene	LOF mutant phenotype	Homologs	Gene product
		HOM-C	
ceh-13	Small, often Unc with abnormal vulva; abnormal larvae	*Drosophila* labial	Homeodomain protein
lin-39	Egl, Muv, Vul. P3.aap to P8.aap (presumptive VC neurons) die in hermaphrodites	*Drosophila* sex combs reduced	Homeodomain protein
mab-5	Posterior midbody cell fates	*Drosophila* Antenopedia	Homeodomain protein
egl-5	Egl, Unc, Mec in posterior	*Drosophila* Abdominal-B	Homeodomain protein
php-3	Wild-type, php-3, nob-1 double is embryonic lethal	*Drosophila* Abdominal-B	Homeodomain protein
nob-1	Wild-type, php-3, nob-1 double is embryonic lethal	*Drosophila* Abdominal-B	Homeodomain protein
		Helix-Loop-Helix	
cnd-1	Unc, ventral nerve cord motor neuron defects	Mammalian NeuroD	Basic HLH
hlh-2	Embryonic lethal by RNAi	MammalianE/*Drosophila* Daughterless	Basic HLH
hlh-3	Egl	*Drosophila* Achaete and Scute	Basic HLH
hlh-4	Unknown	*Drosophila* Achaete and Scute	Basic HLH
hlh-6	Unknown	*Drosophila* Achaete and Scute	Basic HLH
hlh-12	Unknown	*Drosophila* Achaete and Scute	Basic HLH
hlh-14	Vab: posterior defects, Unc: ventral coiling, Egl: loss of PVQ, HSN and PHB neurons	*Drosophila* Achaete and Scute	Basic HLH
lin-22	Lin	*Drosophila* hairy/E(spl)	Basic HLH
lin-32	Lin, many neuroblasts adopt hypodermal fates	*Drosophila* Atonal	Basic HLH
ngn-1	Axon outgrowth and guidance	Mammalian Neurogenin	Basic HLH

(*Continued*)

TABLE I (*Continued*)

Gene	LOF mutant phenotype	Homologs	Gene product
		Touch Neuron Neurogenesis	
ced-3	Ced	Mammalian Caspase2/*Drosophila* Ice	Caspase domain protease
ced-4	Ced	Mammalian apoptotic protease activating factor 1	Caspase recruitment domain
egl-44	Egl, misexpression of touch receptor genes in FLP neurons	Transcription enhancer factor (TEF) proteins	TEA/ATTS domain family transcription factor
egl-46	Egl, misexpression of touch receptor genes in FLP neurons	TFIIA-like zinc finger protein family	C2H2 zinc finger
lin-4	Lin, heterochronic, lack adult cuticle	Micro RNA	Micro RNA
lin-14	Lin, heterochronic lineage alterations	None	Nuclear protein
lin-32	See above		
mec-3	Mec	Human LIM/homeobox protein Lhx5	LIM homeodomain
mec-4	Mec	Amiloride-sensitive sodium channel gamma-subunit	Amiloride-sensitive Na+ channel
mec-7	Mec	Beta-tubulin	Beta-tubulin
pag-3	Unc, ventral nerve cord lineage defects, BDU neurons adopt mechanoreceptor fate	*Drosophila* Senseless and mammalian Gfi-1	C2H2 zinc finger
sem-4	Egl, additional axon and lineage defects	Human Sal-like protein 3	C2H2 zinc finger
unc-86	Unc, Mec, Egl, lethargic other pleotropic effects	Human BRN-3	POU-type homeodomain
vab-15	Vab, Mec; Tab; Unc; Egl	*Drosophila* msh	Homeodomain
		LIM Homeodomain	
ceh-10	Unc, Egl, Muv, withered tail and defects in CAN cell migration	Vertebrate retina protein Chx10 from mice	LIM homeodomain protein
ceh-14	Athermotactic	Human LIM homeobox protein LHX3	LIM homeodomain protein
lim-4	AWB sensory neuron switches fate to AWC-like	Murine Lhx6 and L3/Lhx7, and *Drosophila* ARROWHEAD	LIM homeodomain protein

lim-6	ASEL takes on ASER fate	Human LIM homeobox transcription factor 1, beta	LIM homeodomain protein
lim-7	Unknown	Human LIM homeobox transcription factor Islets	LIM homeodomain protein
lin-11	Unc, Lin	LIM homeobox proteins	LIM homeodomain protein
mec-3	See above		
ttx-3	Cryophilic	Human LIM/homeobox protein Lhx2	LIM homeodomain protein
	Partial Loss of Neuronal Identity		
fax-1	Weak kinker, and defects in axon guidance	Vertebrate photoreceptor-specific nuclear receptor	C4-type zinc finger steroid receptor
unc-4	Unc, VA motoneurons have synaptic inputs appropriate to VB motoneurons	Human homeobox protein ARX	Paired-class homeodomain protein
unc-25	Unc, shrinker, slow, poor backing, small; deficient in GABA	*Drosophila* Gad1	Pyridoxal-dependent decarboxylase
unc-30	Unc, shrinker	Vertebrate Pitx homeodomain proteins	Paired-class homeodomain protein
unc-42	Unc, Egl-c, HSN fasciculation abnormal	Human Pax-7	Paired-class homeodomain protein
unc-47	Unc	Mammalian Vesicular inhibitory amino acid transporter	transmembrane vesicular GABA transporter
zag-1	Misexpression of genes, axonal defects, migration defects	Vertebrate ZEB/*Drosophila* zfh-1	Zinc finger homeodomain protein
	Other Important Genes Not Covered in This Review		
ceh-20	Vul, Egl, SM migration defects. Unc, defective in Q daughter migrations, M lineage. Stronger alleles are larval lethal. Males do not mate.	*Drosophila* Extradenticle (Exd)	PBC class homeodomain protein
ceh-23	sra-11 expression is reduced in adults	*Drosophila* empty spiracles and Distal-less	Homeodomain protein
ceh-36	Loss of neuronal asymmetry	*Drosophila* orthodentical	Homeodomain protein
che-1	Defective chemotaxis to Na	*Drosophila* GLASS	C2H2-type zinc finger
cog-1	Loss of neuronal asymmetry	Human Nkx6	Homeodomain
die-1	Zygotic embryonic lethal	No well-characterized homolog	C2H2 zinc finger protein

(*Continued*)

TABLE I (*Continued*)

Gene	LOF mutant phenotype	Homologs	Gene product
ham-1	Certain asymmetric cell divisions become symmetric	No well-characterized homolog	Novel protein with winged helix DNA binding domain
lin-26	RNAi embryonic lethal. Hypomorphic alleles: P neuroblast and ectodermal defects	Human HKR1 protein	C2H2 zinc finger protein
lin-49	Loss of neuronal asymmetry. Males have abnormal spicules	Human Peregrin	PHD finger domain protein
pal-1	Emb, Vab, V6 lineage transformed into V1-V4	*Drosophila* caudal	Helix-turn-helix and Homeobox domain
unc-119	Unc, Dpy, Egl	Rat retinal protein 4	GMP-PDE delta subunit (fourth subunit of rod-specific cGMP phosphodiesterase)
unc-3	Unc	Immunoglobulin (Ig) domain-containing transcription factors such as OLF-1 or SU(H)	Basic HLH
unc-37	Emb, Unc	*Drosophila* GROUCHO	Transducin-like WD-repeat protein
vab-3	Vab, Lin, disorganized hypodermal and anterior sensory anatomy	Human PAX6	Paired homeodomain
vab-7	Vab, male and hermaphrodite tails deformed, ventral nerve cord defects	*Drosophila* even-skipped	Homeodomain

The phenotypes listed are: Ced = CEll Death abnormality; Dpy = DumPY (shorter than wild-type); Egl = EGg Laying defective; Egl-c = EGg Laying constitutive; Emb = abnormal EMBryogenesis; Lin = abnormal cell LINeage; Mec = MEChanosensory abnormality; Muv = MUltiple Vulva; Tab = Touch ABnormal; Unc = UNCoordinated; Vab = Variable ABnormal morphology; Vul = VULvaless.

single pair of interneurons (Ward *et al.*, 1975). The nervous system contains approximately 5,000 chemical synapses, 600 gap junctions, and 2,000 neuromuscular junctions. The location of chemical synapses is about 75% reproducible from animal to animal (Durbin, 1987).

Information about the functions of many of these neurons has been obtained by a variety of approaches. First, the morphology of many of the neurons has suggested their function (White *et al.*, 1986). For instance, many of the sensory neurons have specialized endings associated with the openings in the cuticle and motor neurons have neuromuscular junctions (NMJs) (Perkins *et al.*, 1986; White *et al.*, 1986). Second, laser killing of single neurons or groups of neurons followed by testing for behavioral or developmental defects has provided functional information (Sulston and White, 1980; Sulston *et al.*, 1983). Finally, the expression pattern and mutant phenotypes of genes have provided insights about specific neurons.

The neurons, like the rest of the cells in *C. elegans*, are generated from an essentially invariant cell lineage. Most neurons are born near where they lie in the adult animal but some migrate and the migrations are also invariant (Sulston and Horvitz, 1977; Sulston *et al.*, 1983).

Often neural subtypes are produced by sets of neuroblasts that undergo stereotyped cell lineages to produce specified groups of cells, each with a defined fate. This is similar to the sensory organ precursor (SOP) sublineages that generate the external sensory organs in *Drosophila* and several neuronal sublineages in mammals such as those that produce the sensory organ patches in mammalian inner ear (Adam *et al.*, 1998).

An example of these sorts of stereotyped cell lineages is the P-neuroblasts that divide to generate the ventral nerve cord motor neurons (Fig. 1). During the first larval stage, 13 neuroblasts arranged along the length of the animal each undergo a stereotyped division pattern to produce five neurons each. The fate of these neurons is similar from one lineage group to another but there is some variation along the length of the animal (Sulston *et al.*, 1983). The correlation of neuronal cell type with lineage pattern suggests that genetic pathways couple neural specification to lineage and the variations on the canonical cell lineage suggest that other genes modify these modular cell and lineage mechanisms (Ruvkun, 1997). One class of genes that modify modular cell and lineage patterns are the homeobox cluster genes (HOM-C genes).

II. HOM-C Genes

As in *Drosophila* and vertebrates, a cluster of homeobox genes regulates spatial patterning along the length of *C. elegans* (Fig. 1). These genes are expressed sequentially in overlapping regions along the length of the body according to

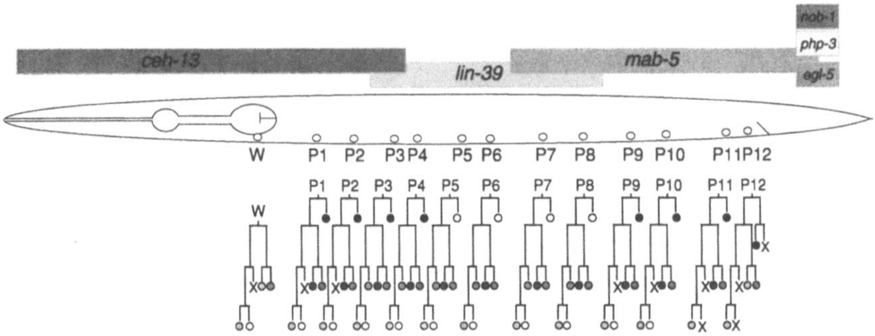

FIG. 1. The HOM-C genes *ceh-13, lin-39, mab-5, egl-5, php-3,* and *nob-1* affect the fate of neurons and other cells along the length of *C. elegans.* The colored boxes represent the approximate regions over which each HOM-C gene is expressed. The worm depicted is at the early first larval (L1) stage. The circles labeled W, P1, P2, and so on represent the P-neuroblasts that undergo the stereotyped cell divisions shown below. While each of the P-neuroblasts undergo a similar set of divisions, there are variations on this theme along the length of the animal and there are differences in the fates of the neurons generated by these lineages. The cell fates are indicated by colored circles as follows: orange are AVF neurons, yellow are VB neurons, green are VA neurons, blue are VD neurons, purple are AS neurons, red are VC neurons, and gray is a PDB neuron. The black cells form hypodermis and the white cells form vulva. Some of the hypodermal cells and all of the vulval cells undergo further cell divisions at the L3 stage that are not shown. (See Color Insert.)

their order on the chromosome (Wang *et al.*, 1993). The homeobox cluster in *C. elegans* has six genes—*ceh-13, lin-39, mab-5, egl-5, php-3* and *nob-1*—which are spread across 5 Mb of chromosome III. These are homologous to labial (*ceh-13*), sex combs reduced (*lin-39*), *fushi tarazu* (*mab-5*), and abdominal-B (*egl-5, php-3,* and *nob-1*) in *Drosophila* (Aboobaker and Blaxter, 2003; Kenyon and Wang, 1991; Ruvkun and Hobert, 1998; Wang *et al.*, 1993). There has been some question as to which homeobox cluster gene *mab-5* corresponds. Aboobaker and Blaxter (2003) examined the homeobox cluster genes in numerous nematodes and concluded that *mab-5* is derived from the same ancestral gene as *fushi tarazu. fushi tarazu* in *Drosophila* has been co-opted for other roles but in many species the *fushi tarazu* ortholog is an active member of the homeobox cluster.

ceh-13 is expressed in the anterior of the animal and affects anterior patterning during embryogenesis. *lin-39* is expressed in the middle of the embryo and affects vulval and neuronal lineages and cell fates in the middle of the animal. *mab-5* is expressed in the anterior half of the animal and its expression overlaps with *lin-39* on one side and *egl-5, php-3,* and *nob-1* on the other. In *mab-5* loss-of-function mutants, neuroblast lineages, and migrations are abnormal. In the male, the tail is grossly abnormal and the V-rays are missing. *egl-5* is expressed in the posterior of the animal and in *egl-5* mutants the fates of cells within the tail region, including the hermaphrodite-specific neuron (HSN) are altered. *php-3* and *nob-1* appear to be a recently duplicated gene pair (Aboobaker and Blaxter, 2003) and both genes are required for normal posterior development. The *C. elegans* homeobox cluster genes interact with each other and with various genes involved in neurogenesis to specify cell fates along the length of the animal.

III. Basic Helix-Loop-Helix Genes

Basic helix-loop-helix (bHLH) proteins are a large family of transcription factors that function in numerous developmental processes including myogenesis, hematopoiesis, cardiogenesis, and neurogenesis (Gering *et al.*, 1998; Hallam *et al.*, 2000; Jennings *et al.*, 1994; Thomas *et al.*, 1998). In many cases, the bHLH proteins appear to have overlapping and redundant functions. bHLH proteins can interact with themselves to form homodimers or with each other to form heterodimers via the HLH domain and they can bind to the deoxyribonucleic acid E-box sequence CANNTG, via their basic domain.

The proneural bHLH proteins in *Drosophila* fall into two major subfamilies: the Achaete-Scute complex family and the Atonal family. Both are expressed transiently in proneural clusters and both interact with a ubiquitous helix-loop-helix protein, Daughterless, to form heterodimers. While the Achaete-Scute complex proteins are involved in the generation of external sensory organs, Atonal is involved in generating two other classes of sensory organs, chordotonal organs and photoreceptors (Jarman *et al.*, 1993, 1994). Another Atonal family member, Amos, is involved in generating a subtype of multiple dendritic organs and olfactory neurons (Huang *et al.*, 2000). Because different proneural genes induce different types of neural structures they must carry information for cell type in addition to being proneural. A fifth *Drosophila* bHLH protein, Asense, is capable of inducing neurogenesis when ectopically expressed, but is expressed later in neurogenesis and has pan-neural functions *in vivo* (Dominguez and Campuzano, 1993; Gonzalez *et al.*, 1998).

The proneural bHLH genes act redundantly and are both necessary and sufficient for external sensory organ development. Thus, in loss-of-function *achaete/scute* double mutants, the cells at the position where the Achaete-Scute proneural clusters normally form adopt an epidermal fate rather than becoming sensory organs. On the other hand, if either *achaete* or *scute* is expressed ectopically, ectopic external sensory organs form. Thus, Achaete and Scute are positive regulators that cause ectodermal cells to become external sensory organs rather than epidermal cells.

The vertebrate proneural genes are divided into four subfamilies: ASH for **A**chaete-**S**cute **h**omology (in mouse the name MASH is used and in *Xenopus*

XASH is used); Atonal related or ATH (in mouse, MATH); Neurogenin; and NeuroD. Members of each family are expressed transiently in overlapping regions and act in a redundant and sequential manner.

The Achaete-Scute, Atonal, and Neurogenin family members have been shown to specify neuroblast cell fate and, therefore, are proneural. For instance, mice lacking MASH1 activity have neurogenesis defects in the ventral telencephalon and olfactory sensory epithelium (Casarosa *et al.*, 1999; Guillemot and Joyner, 1993; Guillemot *et al.*, 1993). In vertebrates, the proneural genes are expressed in subsets of neuroepithelial cells and expression of proneural genes causes these cells to delaminate and generate neurons (Bertrand *et al.*, 2002).

As with the *Drosophila* proneural genes, the vertebrate proneural genes are also involved in neuronal subtype specification in the central nervous system (CNS) and in the peripheral nervous system (PNS) (Goridis and Brunet, 1999; Lo *et al.*, 1998; Tuttle *et al.*, 1999). MASH1 is required for neurogenesis in the hindbrain, ventral forebrain, and olfactory epithelia (Cau *et al.*, 1997; Guillemot and Joyner, 1993; Guillemot *et al.*, 1993; Lo *et al.*, 1991). MATH1 is required for neurogenesis of granule cells in the cerebellum (Ben-Arie *et al.*, 1997; Helms and Johnson, 1998). Neurogenins are required for development of the cranial and dorsal root ganglia (Ben-Arie *et al.*, 1997; Fode *et al.*, 1998; Sommer *et al.*, 1996). The NeuroD subfamily, like Asense in *Drosophila*, can induce neurogenesis in non-neuronal ectoderm when expressed ectopically, but *in vivo*, appears to be expressed later than the proneural genes. For instance, ectopic expression of NeuroD by injection into *Xenopus* oocytes induces non-neuronal ectoderm to differentiate into neurons and causes the premature differentiation of neuroblasts (Lee *et al.*, 1995).

In the mouse, NeuroD is transiently expressed in post-mitotic neurons outside the ventricular zone (Lee, 1997). A knockout of this gene leads to major defects in brain morphology, perhaps due to neuroblasts undergoing massive programmed cell death (Miyata *et al.*, 1999). In the mouse retina, NeuroD is required for cells to become neurons and glia, for interneurons to develop, and for cell survival (Bertrand *et al.*, 2002).

There are approximately 35 bHLH protein-encoding genes in *C. elegans* (Ledent and Vervoort, 2001). As in *Drosophila*, the *C. elegans* neurogenic bHLH proteins can be grouped into two families: the Achaete-Scute complex related and the Atonal related. The Atonal-related bHLH proteins can be further divided into three groups: the Neurogenin group, the NeuroD group, and the Atonal group (Ledent and Vervoort, 2001). In *C. elegans* there are five genes that encode Achaete-Scute related proteins: *hlh-3, hlh-4, hlh-6, hlh-12,* and *hlh-14.* There are three genes that encode Atonal-related proteins: *ngn-1* of the neurogenin group, *cnd-1* of the NeuroD group, and *lin-32* of the Atonal group. Only three of these bHLH proteins have so far been shown to have proneural functions and none appear to be exclusively proneural since all appear to also act later in neurogenesis as well.

A. HLH-2, THE E/DAUGHTERLESS HOMOLOGUE IN *C. ELEGANS*

The *Drosophila* and mammalian proneural bHLH proteins form heterodimers with the ubiquitous E/Daughterless class of HLH proteins and this is also true of the bHLH proneural proteins in *C. elegans* (Frank *et al.*, 2003; Portman and Emmons, 2000; Thellman *et al.*, 2003). *hlh-2* is expressed broadly up to about 350 min after fertilization but then becomes restricted to neurons and neuroblasts (Krause *et al.*, 1997). Knockout of *hlh-2* by antisense ribonucleic acid is lethal with pleiotropic defects (Krause *et al.*, 1997), which may reflect the involvement of HLH-2 in numerous processes.

B. LIN-32

The most extensively characterized bHLH neurogenic gene in *C. elegans* is *lin-32*. As shown in Fig. 2, mutations in *lin-32* cause the transformation of ray neuroblasts in the male tail to epidermal cells, and ectopic expression of *lin-32* results in the formation of additional ray papillae, which shows that LIN-32 is proneural (Zhao and Emmons, 1995). The rays are 18 sensory structures that contain two sensory neurons each in the male tail that help the male sense the hermaphrodite during mating. The lineages that give rise to the rays are similar to the SOP lineages that give rise to external sensory organs in *Drosophila;* these may both derive from an ancestral sublineage (Portman and Emmons, 2000).

LIN-32 forms heterodimers with the E/Daughterless ortholog HLH-2 and this heterodimer can bind to the E-box motif *in vitro* (Portman and Emmons, 2000). Furthermore, partial loss-of-function mutations in *hlh-2* can enhance the ray defects caused by partial loss-of-function mutations in *lin-32* (Portman and Emmons, 2000), which is consistent with them functioning as heterodimers *in vivo*. In *lin-32* loss-of-function mutants, the rays often lack fully differentiated neurons, the neurons lack cognate rays, and the ray neurons have defective processes (Portman and Emmons, 2000). Thus, the LIN-32/HLH-2 heterodimer is involved in specification of multiple cell fates and proper development of differentiated cell characteristics.

C. LIN-22

Hairy and the *Enhancer of split* complex genes inhibit neurogenesis in *Drosophila*. *C. elegans* has a *Hairy/Enhancer of split* homologue, *lin-22*, and as shown in Fig. 2, *lin-22* is also involved in inhibiting neurogenesis of postdeirids and rays (Wrischnik and Kenyon, 1997). The rays and postdeirids arise from lateral epidermal stem cells called V5 and V6. V5 and V6 are the most posterior of a group of six lateral

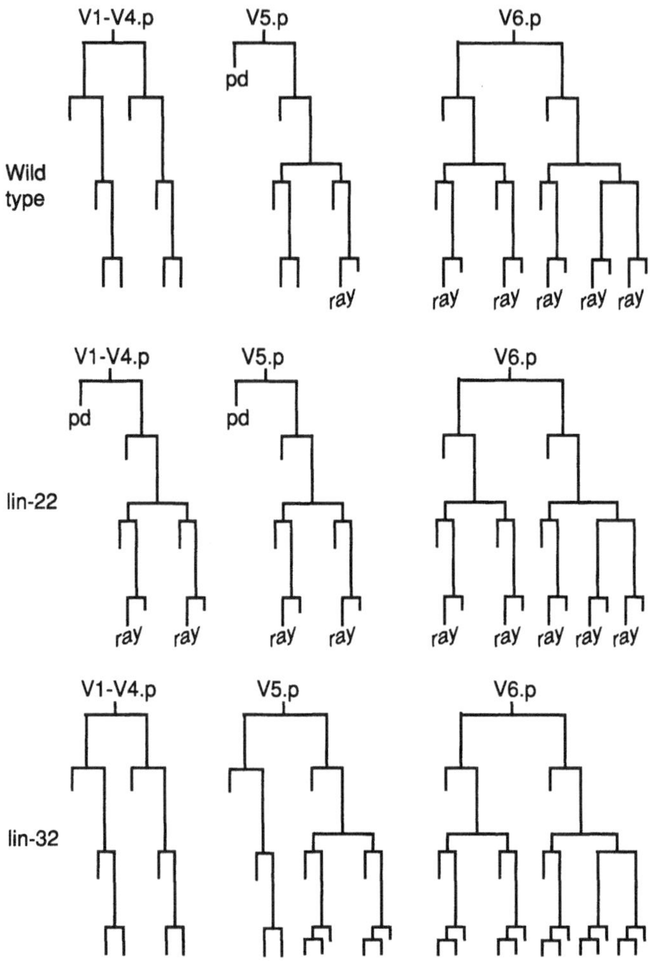

FIG. 2. Lineages of the lateral epidermal stem cells V1 to V6 in wild-type males, *lin*-22 loss-of-function mutants, and *lin*-32 loss-of-function mutants. In wild-type animals, V1 to V4 give rise to only epidermal cells and V5 gives rise to the postdeirid neuron (pd) and sensory neurons in a ray of the male tail. The ray sublineages are marked as "ray." V6 gives rise to five more rays. In *lin*-22 mutants, V1 to V4 give rise to extra postdeirid-like neurons and extra rays. In *lin*-32 mutants, the postdeirids and rays are missing.

epidermal cells named V1 to V6. Normally the anterior cells (V1 to V4) generate only epidermal cells while V5 and V6 give rise to more complex lineages that produce both neural and epidermal cells.

In *lin-22* loss-of-function mutants, the anterior V-cells have V5-like lineages and generate extra postdeirids and ray neuroblasts (Wrischnik and Kenyon,

1997). LIN-22 inhibits these V5-like lineages in V1 to V4 by inhibiting both the HOM-C gene *mab-5* and *lin-32*. Interestingly, another bHLH protein, REF-1, which is related to Hairy but not as closely related as LIN-22, normally prevents the V6 cell from going through V5-like lineages (Alper and Kenyon, 2001).

D. HLH-14

The Achaete-Scute homologue HLH-14 promotes neurogenesis and, like LIN-32 and LIN-22, affects cell lineages. *hlh-14* loss-of-function mutants do not form PVQ, HSN, or PHB neurons because the PVQ/HSN/PHB neuroblast assumes the fate of its sister and gives rise to a hypodermal cell and a T-Blast–like cell instead (Frank *et al.*, 2003).

HLH-14 and HLH-2 appear to act together to specify neuronal lineages and fate. Partial loss-of-function *hlh-2* mutants often have the same lineage defect as *hlh-14* mutants and partial loss-of-function alleles of *hlh-14* are enhanced for the lineage defect by partial loss-of-function alleles of *hlh-2* (Frank *et al.*, 2003). Like *lin-32, hlh-14* has proneural characteristics but is not strictly proneural as it is required for differentiation as well.

E. HLH-3

HLH-3 is an Achaete-Scute homologue that is present in many neuronal precursors (Krause *et al.*, 1997). Its role in neurogenesis is not yet clear. RNAi of *hlh-3* showed that *hlh-3* is required for programmed cell death of the NSM sister cell, which becomes a neuron when programmed cell death is blocked (Thellman *et al.*, 2003). HLH-3 forms heterodimers with HLH-2 and the heterodimers can bind to E-boxes in the *egl-1* gene (Thellman *et al.*, 2003). Activation of *egl-1* induces programmed cell death in the NSM sister cell.

HLH-14, LIN-32, and HLH-3 all affect asymmetric cell lineages. In *hlh-14* mutants the PVQ/HSN/PHB neuroblast is partially transformed into its sister (Frank *et al.*, 2003); in *lin-32* mutants the V5.pa postdeirids neuroblast takes on a V5.pp hypodermal blast fate (Zhao and Emmons, 1995); and in *hlh-3* mutants the NSM sister fails to die and takes on an NSM-like fate (Thellman *et al.*, 2003).

F. NGN-1 AND CND-1

NGN-1 is the only *C. elegans* neurogenin-like protein. To date nothing has been published on *ngn-1*, but the Emmons laboratory found that it has a role in axon outgrowth and guidance (D. Portman, O. Hobert, S. Emmons, 2001, personal communication).

CND-1 is the only *C. elegans* NeuroD homologue. *cnd-1* mutants have fewer motor neurons and those that form are altered in their fate and terminal differentiation characteristics (Hallam *et al.*, 2000). During embryogenesis, *cnd-1::gfp* is expressed early in the AB lineage and expression is maintained in neuronal lineages but then diminishes in most differentiated neurons. During larval development, *cnd-1::gfp* is expressed in the ventral nerve cord lineages and in *cnd-1(ju29)* mutants there are fewer ventral nerve cord neurons at the L1 stage. This suggests that CND-1 is required to maintain mitotic competence in a subset of cells. In mutants, some terminal differentiation markers appeared to be expressed in ventral nerve cord neurons randomly. Furthermore, these mutants showed defects in neuronal terminal differentiation traits such as axonal morphology (Hallam *et al.*, 2000).

CND-1 is most similar to NeuroD and NGN-1 is most similar to neurogenin but both proteins show some hybrid homology to both NeuroD and neurogenin. Furthermore, the expression and function of CND-1 is similar to both NeuroD and neurogenin so these proteins may represent an ancestral form with both NeuroD and Neurogenin like roles (Hallam *et al.*, 2000).

IV. Neurogenesis of the Touch Receptor Neurons

There is too much data on the development of every type of neuron in *C. elegans*, so we will focus on what is perhaps the most extensively characterized type of neuron, the touch receptor neurons (sometimes called the mechanoreceptor neurons). These neurons detect very light touch and their function can be seen if the animal is touched lightly with an eyebrow hair. If touched in the posterior half, *C. elegans* move forward, and if touched in the anterior half, *C. elegans* move backwards (Chalfie and Sulston, 1981; Sulston *et al.*, 1975). Thousands of mutations have been isolated that affect the development or function of the touch receptor neurons by screening for animals that do not respond to the touch assay or by looking for mutants that fail to express or misexpress touch receptor neuron-specific genes (Chalfie and Au, 1989; Chalfie and Sulston, 1981; Chalfie and Thomson, 1982; Chalfie *et al.*, 1981; Du and Chalfie, 2001; Mitani *et al.*, 1993; Sulston *et al.*, 1975). Because many alleles have been identified in each gene required for touch sensitivity, the eyebrow hair screen is very likely saturated. On the other hand, screens for mutation that affect touch receptor neuron-specific gene expression are unlikely to be saturated (Du and Chalfie, 2001) and these screens are capable of identifying mutations that result in severe phenotypes.

A. MEC-3

As shown in Fig. 3, the most proximal regulator of touch receptor neuron gene expression is *mec-3*. *mec-3* is a neuronal cell-type selector gene because in animals with loss-of-function mutations in *mec-3*, the cells that would have formed

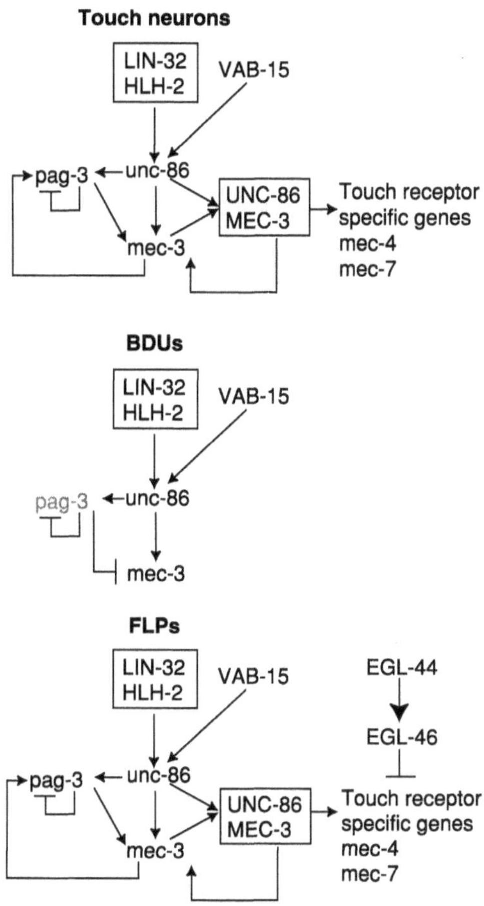

FIG. 3. Transcriptional networks regulating touch neuron development. In the touch receptor neurons, a heterodimer formed from the bHLH transcription factors LIN-32 and HLH-2 together with VAB-15 activate unc-86. UNC-86 then activates mec-3. UNC-86 and MEC-3 form a heterodimer that further activates mec-3 and activates the touch receptor neuron-specific genes such as mec-4 and mec-7. pag-3 is activated by UNC-86 and MEC-3 and further activates mec-3. In the BDU neurons, pag-3 is expressed at low levels and represses mec-3, which prevents the BDUs from becoming touch receptor neurons. In the FLPs, mec-3 is expressed, but the touch receptor-specific genes are repressed by egl-46.

touch receptor neurons still become neurons but not touch receptors. *mec-3* encodes a LIM-homeodomain protein that is expressed in eight cells including the six touch receptor neurons, the two FLP neurons that detect touch to the tip of the head, and the two PVD neurons that detect harsh touch such as touch with a wire (Way and Chalfie, 1989).

mec-3 is activated by UNC-86, a POU-homeodomain protein. UNC-86 activates *mec-3* expression and once MEC-3 protein is made, UNC-86 and MEC-3 form a heterodimer that further activates *mec-3*, and also activates the expression of touch receptor neuron-specific genes such as *mec-7* and *mec-4* (Duggan *et al.*, 1998; Lichtsteiner and Tjian, 1995; Xue *et al.*, 1992). MEC-17 is required to maintain *mec-3* expression in late larvae and adults (Mitani *et al.*, 1993; Way and Chalfie, 1989).

B. UNC-86

unc-86 is expressed in 47 neuronal lineages and in 57 postmitotic neurons including all the cells that express *mec-3* (Finney and Ruvkun, 1990). UNC-86 specifies many neuroblast and neuronal fates and loss-of-function mutations in *unc-86* lead to neuroblast lineages where the daughter cells retain the identity of their mothers so the mother's lineage is reiterated (Chalfie *et al.*, 1981). This leads to the loss and duplication of neurons and behavioral abnormalities in mechanosensation, egg laying, chemotaxis, and thermotaxis (Mori and Ohshima, 1995).

unc-86 acts at several steps in touch receptor neuron differentiation. It is required to specify lineages that give rise to touch receptor neurons, it is involved in touch receptor neuron specification by activating *mec-3* and it is involved in touch receptor neuron differentiation and function by activating touch receptor neuron-specific genes.

C. LIN-32 AND VAB-15

Two other genes required to specify lineages that generate touch receptor neurons are *lin-32* and *vab-15*. *lin-32* mutants lack the AVM, PVM, and two PLM touch receptor neurons due to lineage defects and the remaining two touch receptor neurons, the ALMs, often fail to migrate to their correct positions (Du and Chalfie, 2001). VAB-15 is an msh-type homeodomain transcription factor. The touch receptor neuron defects in *vab-15* mutants are very similar to those in *lin-32* mutants: the AVM, PVM, and two PLMs are missing and the ALMs often fail to migrate to their correct positions (Du and Chalfie, 2001). In *vab-15 lin-32* double mutants, the ALMs are also often missing, which suggests that *vab-15* and *lin-32* act redundantly (Du and Chalfie, 2001).

D. LIN-4 AND LIN-14

Two more genes that control lineages that generate touch receptor neurons are *lin-4* and *lin-14*. *lin-4* and *lin-14* are heterochronic genes that regulate the timing of lineages. In *lin-14* gain-of-function mutants, there is repetition of several lineages that occur during the first larval stage, while in *lin-14* loss-of-function mutants, lineages that occur after the first larval stage occur prematurely. *lin-4* represses *lin-14* (Ambros, 1989; Arasu *et al.*, 1991) so *lin-4* loss-of-function mutations are similar to *lin-14* gain of function mutations (Ambros and Horvitz, 1984; Chalfie *et al.*, 1981). The AVM and PVM touch receptor neurons are generated in the first larval stage and the PVD neurons are generated in the second larval stage (Sulston and Horvitz, 1977). *lin-14* loss-of-function mutants lack AVM and PVM touch receptor cells and the AVM/PVM precursor follows a PVD-like lineage. In *lin-14,* gain-of-function mutants the PVD lineage is sometimes converted to an AVM/PVM-like lineage so extra touch cells are generated (Mitani *et al.*, 1993).

lin-4 encodes a conserved non-coding micro-RNA (Lee *et al.*, 1993). While it is not yet clear how important micro-RNAs are for neurogenesis, Johnston and Hobert (2003) identified another miRNA that regulates the asymmetry of the two ASE taste receptor neurons.

E. GENES THAT RESTRICT TOUCH RECEPTOR NEURONS

Several genes in addition to *lin-4* restrict the number of touch cells to six. These include the transcription factor genes *egl-44, egl-46, sem-4,* and *pag-3* (Basson and Horvitz, 1996; Jia *et al.*, 1996; Mitani *et al.*, 1993; Toker *et al.*, 2003; Wu *et al.*, 2001) and the cell death genes *ced-3* and *ced-4,* which allow the death of four cells that become touch receptor neurons if cell death is prevented (Mitani *et al.*, 1993).

egl-44 encodes a transcription enhancer factor family protein and *egl-46* encodes a zinc finger protein (Wu *et al.*, 2001). In *egl-44* and *egl-46* loss-of-function mutants, the FLP neurons express touch receptor neuron genes. In wild-type animals, *mec-3* and *unc-86* are both expressed in the FLPs so *egl-44* and *egl-46* act downstream of *mec-3* and repress touch receptor neuron-specific genes in the FLPs (Wu *et al.*, 2001). In the FLPs (but not in the HSNs) the expression of *egl-46* depends on *egl-44*. In the touch receptor neurons, *egl-46* is expressed independent of *egl-44* but ectopic expression of *egl-44* in the touch receptor neurons inhibits the touch cell fate in an *egl-46*–dependent manner (Wu *et al.*, 2001).

sem-4 encodes a spalt-like zinc finger transcription factor. In *sem-4* loss-of-function mutants the T.pp lineage, which normally gives rise to PVW, PHC, and PLN neurons, is sometimes converted to a PLM precursor-like lineage and the cell that would normally become a PHC becomes a touch receptor neuron. Alternatively, in a smaller percentage of animals, the T.pp lineage resembles the QL lineage, which gives rise to the PVM touch receptor neuron (Toker *et al.*, 2003). SEM-4 binds to a motif in the *mec-3* and *egl-5* genes and repression of these genes represses the touch receptor neuron fate (Toker *et al.*, 2003).

pag-3 encodes a C2H2 zinc finger transcription factor, which is the worm ortholog of *Drosophila Senseless*. Senseless is required for external sensory organ formation and when ectopically expressed, induces ectopic external sensory organs (Nolo *et al.*, 2000). Proneural proteins activate both Senseless and *E(spl)-C* genes. Where Delta has activated Notch, the Notch intracellular domain together with Su(H) further activate the *E(spl)-C* genes, which suppress both Senseless and the proneural genes. Senseless, at a low level of expression, further represses the proneural genes.

In a cell destined to become a sensory organ precursor, the *E(spl)-C* genes are suppressed rather than activated by Su(H). This allows Senseless levels to rise, and once they pass a threshold, Senseless shifts from repression to activation (Jafar-Nejad *et al.*, 2003; Nolo *et al.*, 2000). Senseless thereby acts as a binary switch regulating SOP cell fate (Jafar-Nejad *et al.*, 2003).

In *pag-3* loss-of-function mutants, the BDU interneurons, which are the lineal sisters to the ALM touch receptor neurons also adopt a touch receptor neuron fate (Jia *et al.*, 1996). In the BDU neurons, *pag-3* is expressed at a low level and inhibits *mec-3* expression and in the touch receptor neurons, *pag-3* is expressed at a high level and activates *mec-3* expression (Toms and Aamodt, 2004, unpublished data). *pag-3* also regulates *unc-86* expression in the ventral nerve cord. In the P.n neuroblast lineages, *pag-3* represses *unc-86* expression (Toms and Aamodt, unpublished data) and in *pag-3* loss-of-function mutants the Pn.aaa cells often maintain a Pn.aa identity, and thus reiterate the Pn.aa cell lineage (Cameron *et al.*, 2002), a phenotype similar to that caused by *unc-86* loss-of-function mutations in other lineages.

Each of the genes that either activates or represses the touch cell fate is expressed in many types of neurons and none are expressed specifically in touch receptor neurons. Thus, the touch receptor neuron fate results from the combinatorial action of many genes (Du and Chalfie, 2001; Mitani *et al.*, 1993; Toker *et al.*, 2003; Wu *et al.*, 2001). Furthermore, the transcriptional regulatory networks regulating touch receptor neuron fate may be different in different touch receptor neurons.

Genes have been shown to act at each step in touch receptor neuron specification and differentiation, but few of the genes characterized so far act specifically at a particular step. As we saw for *lin-32* in the male tail, the genes that specify touch receptor neurons usually act at multiple steps. An obvious exception to this is *mec-3*, which is expressed after the last cell division and gives the touch neurons their specific character. As described in the next section, other LIM homeodomain proteins appear to have a similar role in other classes of neurons.

V. The LIM Hypothesis

The LIM-homeobox consists of two zinc finger-like LIM domains, which are protein interaction domains, and a DNA-binding homeodomain. MEC-3 was the first LIM homeodomain protein described and is still the most studied.

C. elegans contains seven LIM homeobox proteins that fall into six families (Hobert and Ruvkun, 1998; Hobert and Westphal, 2000): TTX-3 is in the apterous group, LIM-4 is in the LHX6/7 group, LIM-7 is in the ISLETS group, LIM-6 is in the LMX group, CEH-14 is in the LIM-3 group, and MEC-3 and LIN-11 are both in the LIN-11 group (Hobert and Westphal, 2000).

All of the *C. elegans* LIM-homeobox genes appear to be expressed after the last cell division and to specify subtypes of neurons (Hobert and Ruvkun, 1998; Hobert and Westphal, 2000). For instance, *mec-3* is expressed postmitotically in the touch receptor neurons and loss-of-function mutations cause the cells that would have become touch receptor neurons to become interneurons-like cells. They fail to express touch receptor neuron-specific genes, to migrate correctly, and to elaborate the correct axonal morphology (Way and Chalfie, 1988). The PVD harsh touch sensory neurons of *mec-3* mutants fail to form the elaborate dendritic morphology present in wild-type animals and have defects in harsh touch sensation (Tsalik *et al.*, 2003; Way and Chalfie, 1989). *ttx3, lin-11, lim-4,* and *lim-6,* similar to *mec-3,* are first expressed after the final cell division and their expression is maintained throughout adulthood.

Three separate LIM-homeodomain proteins specify three key thermosensory neurons: the AFD neuron requires CEH-14, the AIY neuron requires TTX-3, and the AIZ neuron requires LIN-11 (Andachi *et al.*, 2001; Cassata *et al.*, 2000; Hobert *et al.*, 1998). The neural thermoregulatory pathway in *C. elegans* includes the AFD, AIY, and AIZ neurons. AFD is the sensory cell for low temperature and it synapses onto AIY. If either AFD or AIY are ablated the animals become unresponsive to low temperature. If AIZ is ablated the animals respond poorly to high temperature (Mori and Ohshima, 1995).

The LIM homeobox gene *ceh-14* is expressed in the AFD neurons. In *ceh-14* loss-of-function mutants, the AFD neurons are correctly placed but fail to function so the animals are athermotactic. Ectopic expression of the wild-type *ceh-14* gene in the AWA and AWB chemosensory neurons of a *ceh-14* mutant restores thermosensitivity, which shows that *ceh-14* is both necessary and sufficient to make certain neurons thermosensory (Cassata *et al.*, 2000).

TTX-3 and LIN-11 are in the AIY and AIZ interneurons, respectively, and each is required for the thermoregulation function of these neurons. *ttx-3* loss-of-function mutants have thermophilic defects and *lin-11* loss-of-function mutants have cryophilic defects (Altun-Gultekin *et al.*, 2001; Hobert *et al.*, 1998). TTX-3 and the paired-type homeodomain protein CEH-10 together regulate another homeobox gene, *ceh-23*, and TTX-3 and CEH-10 regulate all the known subtype features of AIY (Altun-Gultekin *et al.*, 2001).

LIN-11 is also in the ventral nerve cord where it is required for axon bundle fasciculation (Hobert *et al.*, 1998) and in the AWA olfactory neuron where it regulates the AWA-specific gene *odr-7* (Sarafi-Reinach *et al.*, 2001). ODR-7 suppresses AWC fate and activates AWA fate (Sagasti *et al.*, 1999; Sengupta *et al.*, 1994).

Another terminal differentiation feature of neurons is the neurotransmitter receptors they produce. Tsalik et al. (2003) noticed that expression of specific biogenic amine receptors overlaps with LIM-homeobox gene expression, and they found that expression of the biogenic amine receptors sometimes depends on the LIM homeobox genes. *ttx-3* is required for *ser-2* expression in AIY, *lim-4* is required for expression of one splice variant of *ser-2* in RID, and *lim-6* is required for proper *dop-1* and *ser-4* expression in the RIS neurons. LIM homeodomain genes, however, did not affect biogenic amine receptor expression in several other neuron types.

VI. Mutations That Cause the Partial Loss of Neuronal Identity

The genes that we have so far considered orchestrate neurogenesis, or neuronal subtype specification, or both. Some of the LIM-homeodomain proteins may regulate just a subset of the genes required for making specific types of neurons. Other examples of genes that regulate a subset of the genes required for making a specific type of neuron include *zag-1, unc-42, fax-1, unc-4,* and *unc-30.*

zag-1 is a deltaEF1/ZFH-1 zinc-finger homeodomain transcription factor. *zag-1* mutations cause misexpression of neuron-specific genes, block axon branching, interfere with neuronal migrations, and cause axonal guidance errors. *unc-42* mutations affect the expression of glutamate-receptor subunits in a subset of interneurons (Baran *et al.*, 1999). *fax-1* is a nuclear hormone receptor that is required for correct axon path finding and expression of a peptide neurotransmitter precursor (Much *et al.*, 2000). *unc-4* is a homeodomain transcription factor that is involved in defining synaptic input for one class of motor neurons. In *unc-4* mutants, the VA motor neurons receive inputs from neurons that would normally contact the VB motor neurons (Miller and Niemeyer, 1995; Miller *et al.*, 1992; White *et al.*, 1992). UNC-30 is a paired-like homeodomain transcription factor—orthologous to vertebrate Pitx homeodomain proteins—that controls the differentiation of all type D GABA-ergic motor neurons by regulating the expression of *unc-25* and *unc-47*, whose products are responsible for GABA formation and secretion, respectively.

VII. Conclusions

Neurogenesis in nematodes, insects, and mammals has much in common. In each system, bHLH transcription factors have a proneural role. In *C. elegans* this role is coupled to the regulation of lineages but not to Notch signaling, as it is in *Drosophila*. In all of these animals, the proneural genes also have roles later in neurogenesis.

In *C. elegans*, most of the genes involved in neurogenesis appear to act at multiple levels. For instance, a gene-like *unc-86* controls lineages, is required for neurogenesis, activates the subtype selector gene *mec-3*, and together with *mec-3* activates the genes that build the touch receptor neurons, such as the ion channel gene *mec-4* and the β-tubulin gene *mec-7*. An important exception is the LIM-homeodomain transcription factors. They are expressed after the last cell division and regulate neuronal subtype-specific genes. There are also genes that regulate a subset of the genes required in specific types of neurons.

The logic of the transcriptional regulatory networks regulating neurogenesis in *C. elegans* is beginning to emerge. In the touch neurons, *unc-86* and *mec-3* are part of a feed-forward loop. Feed-forward loops of this type delay the dynamic response of a circuit. *mec-3* feeds back on its self forming a positive feedback loop, and *mec-3* activates *pag-3*, which activates *mec-3* forming a second, redundant positive feedback loop. These positive feedback loops lock the cells into a touch receptor neuron fate. Like Senseless, PAG-3 appears to act as a bipolar switch and when expressed at low levels in the BDU neurons PAG-3 represses *mec-3* and when expressed at high levels in the touch neurons, PAG-3 activates *mec-3*.

References

Aboobaker, A., and Blaxter, M. (2003). Hox gene loss during dynamic evolution of the nematode cluster. *Curr. Biol.* **13,** 37–40.

Adam, J., Myat, A., LeRoux, I., Edison, M., Henrique, D., Ish-Horowicz, D., and Lewis, J. (1998). Cell fate choice and the expression of Notch, Delta and Serrate homologues in the chick ear: Parallels with *Drosophila* sense-organ development. *Development* **125,** 4645–4654.

Albertson, D., and Thomson, J. (1976). The pharynx of *C. elegans*. *Philos. Trans R Soc. Lond. B Biol. Sci.* **275,** 299–325.

Alper, S., and Kenyon, C. (2001). REF-1, a protein with two bHLH domains, alters the pattern of cell fusion in *C. elegans* by regulating Hox protein activity. *Development* **128,** 1793–1804.

Altun-Gultekin, Z., Andachi, Y., Tsalik, E., Pilgrim, D., Kohara, Y., and Hobert, O. (2001). A regulatory cascade of three homeogox genes, *che-10, ttx-3* and *ceh-23,* controls cell fate specification of a defined interneuron class in *C. elegans. Development* **128,** 1951–1969.

Ambros, V. (1989). A hierarchy of regulatory genes controls a larva-to-adult developmental switch in *C. elegans. Cell* **57,** 49–57.

Ambros, V., and Horvitz, H. (1984). Heterochronic mutants of the nematode *C. elegans. Science* **226,** 409–416.

Andachi, Y., Tsalik, E. L., Pilgrim, D., Kohara, Y., Hobert, O., and Altun-Gultekin, Z. (2001). A regulatory cascade of three homeobox genes, *ceh-10, ttx-3* and *ceh-23,* controls cell fate specification of a defined interneuron class in *C. elegans. Development* **128,** 1951–1969.

Arasu, P., Wightman, B., and Ruvkun, G. (1991). Temporal regulation of *lin-14* by the antagonistic action of two other heterochronic genes, *lin-4* and *lin-28. Genes Dev.* **5,** 1825–1833.

Baran, R., Aronoff, R., and Garriga, G. (1999). The *C. elegans* homeodomain gene *unc-42* regulates chemosensory and glutamate receptor expression. *Development* **126,** 2241–2251.

Basson, M., and Horvitz, H. (1996). The *Caenorhabditis elegans* gene *sem-4* controls neuronal and mesodermal cell development and encodes a zinc finger protein. *Genes Dev.* **10,** 1953–1965.

Ben-Arie, N., Bellen, H., Armstrong, D., McCall, A., Gordadze, P., Guo, Q., Matzuk, M., and Zoghbi, H. (1997). Math1 is essential for genesis of cerebellar granule neurons. *Nature* **390,** 169–172.

Bertrand, N., Castro, D. S., and Guillemot, F. (2002). Proneural genes and the specification of neural cell types. *Nat. Rev. Neurosci.* **3,** 517–530.

Cameron, S., Clark, S. G., McDermott, J. B., Aamodt, E., and Horvitz, H. R. (2002). PAG-3, a Zn-finger transcription factor, determines neuroblast fate in C. elegans. *Development* **129,** 1763–1774.

Casarosa, S., Fode, C., and Guillemot, F. (1999). Mash1 regulates neurogenesis in the ventral telencephalon. *Development* **126,** 525–534.

Cassata, G., Kagoshima, H., Andachi, Y., Kohara, Y., Durrenberger, M., Hall, D., and Burglin, T. (2000). The LIM homeobox gene *ceh-14* confers thermosensory function to the AFD neurons in *Caenorhabditis elegans. Neuron* **25,** 587–597.

Cau, E., Gradwohl, G., Fode, C., and Guillemot, F. (1997). Mash1 activates a cascade of bHLH regulators in olfactory neuron progenitors. *Development* **124,** 1611–1621.

Chalfie, M., and Au, M. (1989). Genetic control of differentiation of the *Caenorhabditis elegans* touch receptor neurons. *Science* **243,** 1027–1033.

Chalfie, M., Horvitz, H. R., and Sulston, J. E. (1981). Mutations that lead to reiterations of the cell lineages of *C. elegans. Cell* **24,** 59–69.

Chalfie, M., and Sulston, J. (1981). Developmental genetics of the mechanosensory neurons of the nematode *Caenorhabditis elegans. Dev. Biol.* **82,** 358–370.

Chalfie, M., and Thomson, J. N. (1982). Structural and functional diversity in the neuronal microtubules of *Caenorhabditis elegans. J. Cell Biol.* **93,** 15–23.

Dominguez, M., and Campuzano, S. (1993). *asense,* a member of the *Drosophila* achaete-scute complex, is a proneural and neural differentiation gene. *EMBO J.* **12,** 2049–2060.

Du, H., and Chalfie, M. (2001). Genes regulating touch cell development in *Caenorhabditis elegans. Genetics* **158,** 197–207.

Duggan, A., Ma, C., and Chalfie, M. (1998). Regulation of touch receptor differentiation by the *Caenorhabditis elegans mec-3* and *unc-86* genes. *Development* **125,** 4107–4119.

Durbin, R. (1987). Studies on the development and organisation of the nervous system of *Caenorhabditis elegans.* Ph.D. dissertation. University of Cambridge, Cambridge, England.

Finney, M., and Ruvkun, G. (1990). The *unc-86* gene product couples cell lineage and cell identity in *C. elegans. Cell* **63,** 895–905.

Fode, C., Gradwohl, G., Morin, X., Dierich, A., LeMeur, M., Goridis, C., and Guillemot, F. (1998). The bHLH protein NEUROGENIN 2 is a determination factor for epibranchial placode-derived sensory neurons. *Neuron* **20,** 483–494.

Frank, C., Baum, F., and Garriga, G. (2003). HLH-14 is a *C. elegans* Achate-Scute protein that promotes neurogenesis through asymmetric cell division. *Development* **130,** 6507–6518.

Gering, M., Rodaway, A., Gottgens, B., Patient, R., and Green, A. (1998). The SCL gene specifies haemangioblast development from early mesoderm. *EMBO J.* **17,** 4029–4045.

Gonzalez, F., Romani, S., Cubas, P., Modolell, J., and Campuzano, S. (1998). Molecular analysis of the *asense* gene, a member of the achaete-scute complex of *Drosophila melanogaster*, and its novel role in optic lobe development. *EMBO J.* **8,** 3553–3562.

Goridis, C., and Brunet, J. (1999). Transcriptional control of neurotransmitter phenotype. *Curr. Opin. Neurobiol.* **9,** 47–53.

Guillemot, F., and Joyner, A. (1993). Dynamic expression for the murine Acheate-Scute homologue Mash-1 in the developing nervous system. *Mech. Dev.* **42,** 171–185.

Guillemot, F., LO, L., Johnson, J., Aurbach, A., Anderson, D., and Joyner, A. (1993). Mammalian achaete-scute homolog 1 is required for the early development of olfactory and autonomic neurons. *Cell* **75,** 463–476.

Hallam, S., Singer, E., Waring, D., and Jin, Y. (2000). The *C. elegans NeuroD* homolog *cnd-1* functions in multiple aspects of motor neuron fate specification. *Development* **127,** 4239–4252.

Helms, A., and Johnson, J. (1998). Progenitors of dorsal commissural interneurons are defined by MATH1 expression. *Development* **125,** 919–928.

Hobert, O., D'Alberti, T., Liu, Y., and Ruvkun, G. (1998). Control of neural development and function in a thermoregulatory network by the LIM homeobox gene *lin-11*. *J. Neurosci.* **18,** 2084–2096.

Hobert, O., and Ruvkun, G. (1998). A common theme for LIM homeobox gene function across phylogeny? *Biol. Bull.* **195,** 377–380.

Hobert, O., and Westphal, H. (2000). Functions of LIM-homeobox genes. *Trends Genet.* **16,** 75–83.

Huang, M., Hsu, C., and Chien, C. (2000). The proneural gene *amos* promotes multiple dendritic neuron formation in the *Drosophila* peripheral nervous system. *Neuron* **25,** 57–67.

Jafar-Nejad, H., Acar, M., Nolo, R., Lacin, H., Pan, H., Parkhurst, S. M., and Bellen, H. J. (2003). Senseless acts as a binary switch during sensory organ precursor selection. *Genes Dev.* **17,** 2966–2978.

Jarman, A., Grau, Y., Jan, L., and Jan, Y. (1993). *Atonal* is a proneural gene that directs chordotonal organ formation in the *Drosophila* peripheral nervous system. *Cell* **73,** 1307–1321.

Jarman, A., Grell, E., Ackerman, L., Jan, L., and Jan, Y. (1994). *Atonal* is the proneural gene for *Drosophila* photoreceptors. *Nature* **369,** 398–400.

Jennings, B., Preiss, A., Delidakis, C., and Bray, S. (1994). The Notch signaling pathway is required for Enhancer of split bHLH protein expression during neurogenesis in the *Drosophila* embryo. *Development* **120,** 3537–3548.

Jia, Y., Xie, G., and Aamodt, E. J. (1996). *pag-3*, a *C. elegans* gene involved in touch neuron gene expression and coordinated movement. *Genetics* **142,** 141–147.

Johnston, R., and Hobert, O. (2003). A microRNA controlling left/right neuronal asymmetry in *Caenorhabditis elegans*. *Nature* **426,** 845–849.

Kenyon, C., and Wang, B. (1991). A cluster of antennapedia-class homeobox genes in a nonsegmented animal. *Science* **253,** 516–517.

Krause, M., Park, M., Zhang, J.-M., Yuan, J., Harfe, B., Xu, S., Greenwald, I., Cole, M., Paterson, B., and Fire, A. (1997). A *C. elegans* E/Daughterless bHLH protein marks neuronal but not striated muscle development. *Development* **124,** 2179–2189.

Ledent, V., and Vervoort, M. (2001). The basic helix-loop-helix protein family: Comparative genomics and phylogenetic analysis. *Genome Res.* **11,** 754–770.

Lee, J. (1997). NeuroD and neurogenesis. *Dev. Neurosci.* **19,** 27–32.

Lee, J., Hollenberg, S., Snider, L., Turner, D., Lipnick, N., and Weintraub, H. (1995). Conversion of *Xenopus* ectoderm into neurons by NeuroD, a basic helix-loop-helix protein. *Science* **268,** 836–844.

Lee, R., Feinbaum, R., and Ambros, V. (1993). The *C. elegans* heterochronic gene *lin-4* encodes small RNAs with antisense complementarity to *lin-14. Cell* **75 5,** 843–854.

Lichtsteiner, S., and Tjian, R. (1995). Synergistic activation of transcription by UNC-86 and MEC-3 in *Caenorhabditis elegans* embryo extracts. *EMBO J.* **14,** 3937–3945.

Lo, L., Johnson, J., Wuenschell, C., Saito, T., and Anderson, D. (1991). Mammalian achaete-scute homolog 1 is transiently expressed by spatially restricted subsets of early neuroepithelial and neural crest cells. *Genes Dev.* **5,** 1524–1537.

Lo, L., Tiveron, M., and Anderson, D. (1998). MASH1 activates expression of the paired homeodomain transcription factor Phox2a, and couples pan-neuronal and subtype-specific components of autonomic neuronal identity. *Development* **125,** 609–620.

Miller, D., and Niemeyer, C. (1995). Expression of the *unc-4* homeoprotein in *Caenorhabditis elegans* motor neurons specifies presynaptic input. *Development* **121,** 2877–2886.

Miller, D., Shen, M., Shamu, C., Burglin, T., Ruvkun, G., Dubois, M., McGhee, M., and Wilson, L. (1992). *C. elegans unc-4* gene encodes a homeodomain protein that determines the pattern of synaptic input to specific motor neurons. *Nature* **355,** 841–845.

Mitani, S., Du, H., Hall, D. H., Driscoll, M., and Chalfie, M. (1993). Combinatorial control of touch receptor neuron expression in *Caenorhabditis elegans. Development* **119,** 773–783.

Miyata, T., Maeda, T., and Lee, J. (1999). NeuroD is required for differentiation of the granule cells in the cerebellum and hippocampus. *Genes Dev.* **13,** 1647–1652.

Mori, I., and Ohshima, Y. (1995). Neural regulation of thermotaxis in *Caenorhabditis elegans. Nature* **376,** 344–348.

Much, J., Slade, D., Klampert, K., Garriga, G., and Wightman, B. (2000). The fax-1 nuclear hormone receptor regulates axon pathfinding and neurotransmission. *Development* **127,** 703–712.

Nolo, R., Abbott, L. A., and Bellen, H. J. (2000). Senseless, a Zn finger transcription factor, is necessary and sufficient for sensory organ development in *Drosophila. Cell* **102,** 349–362.

Perkins, L., Hedgecock, E., Thomson, J., and Culotti, J. (1986). Mutant sensory cilia in the nematode *C. elegans. Dev. Biol.* **117,** 456–487.

Portman, D. S., and Emmons, S. W. (2000). The basic helix-loop-helix transcription factors LIN-32 and HLH-2 function together in multiple steps of a *C. elegans* neuronal sublineage. *Development* **127,** 5415–5426.

Ruvkun, G. (1997). Patterning the Nervous System. *In* "C. elegans II" (D. Riddle, T. Blumenthal, B. Meyer, and J. Priess, Eds.), pp. 543–581. Cold Spring Harbor Laboratory Press, Cold Spring Harbor.

Ruvkun, G., and Hobert, O. (1998). The taxonomy of developmental control in *Caenorhabditis elegans. Science* **282,** 2033–2041.

Sagasti, A., Hobert, O., Troemel, E., Ruvkun, G., and Bargmann, C. (1999). Alternative olfactory neuron fates are specified by the LIM homeobox gene *lim-4. Genes Dev.* **13,** 1794–1806.

Sarafi-Reinach, T., Melkman, T., Hobert, O., and Sengupta, P. (2001). The *lin-11* LIM homeobox gene specifies olfactory and chemosensory neuron fates in *C. elegans. Development* **128,** 3269–3281.

Sengupta, P., Colbert, H., and Bargmann, C. (1994). The *C. elegans* gene *odr-7* encodes an olfactory-specific member of the nuclear receptor superfamily. *Cell* **79,** 971–980.

Sommer, L., Ma, Q., and Anderson, D. (1996). Neurogenins, a novel family of atonal-related bHLH transcription factors, are putative mammalian neuronal determination genes that reveal progenitor cell heterogeneity in the developing CNS and PNS. *Mol. Cell. Neurosci.* **8,** 221–241.

Sulston, J., Dew, M., and Brenner, S. (1975). Dopaminergic neurons in the nematode *C. elegans. J. Comp. Neurol.* **163,** 215–226.

Sulston, J., and White, J. (1980). Regulation and cell autonomy during postembryonic development of *C. elegans*. *Dev. Biol.* **78,** 577–597.

Sulston, J. E., and Horvitz, H. R. (1977). Post-embryonic cell lineages of the nematode *Caenorhabditis elegans*. *Dev. Biol.* **56,** 110–156.

Sulston, J. E., Schierenberg, E., White, J. G., and Thomson, J. N. (1983). The embryonic cell lineage of the nematode *Caenorhabditis elegans*. *Dev. Biol.* **100,** 64–119.

Thellman, M., Hatzold, J., and Conradt, B. (2003). The Snail-like CES-1 protein of *C. elegans* can block the expression of the BH3-only cell-death activator gene *egl-1* by antagonizing the functions of bHLH proteins. *Development* **130,** 4057–4071.

Thomas, T., Yamagishi, H., Overbeek, P., Olson, E., and Srivastava, D. (1998). The bHLH factors, dHAND and eHAND, specify pulmonary and systemic cardiac ventricles independent of left-right sidedness. *Dev. Biol.* **196,** 228–236.

Toker, A., Teng, Y., Ferreira, H., Emmons, S., and Chalfie, M. (2003). The *Caenorhabditis elegans* spalt-like gene *sem-4* restricts touch cell fate by repressing the selector Hox gene *egl-5* and the effector gene *mec-3*. *Development* **130,** 3831–3840.

Tsalik, E., Niacaris, T., Wenick, A., Pau, K., Avery, L., and Hobert, O. (2003). LIM homeobox gene-dependent expression of biogenic amine receptors in restricted regions of the *C. elegans* nervous system. *Dev. Biol.* **263,** 81–102.

Tuttle, R., Nakagawa, Y., Johnson, J., and O'Leary, D. (1999). Defects in the thalamocortical axon pathfinding correlate with altered cell domains in Mash-1-deficient mice. *Development* **126,** 1903–1916.

Wang, B., Muller-Immergluck, M., Austin, J., Robinson, N., Chisholm, A., and Kenyon, C. (1993). A homeotic gene cluster patterns the anterioposterior body axis on *C. elegans*. *Cell* **74,** 29–42.

Ward, S., Thomson, J., White, J., and Brenner, S. (1975). Electron microscopical reconstruction of the anterior sensory anatomy of the nematode *C. elegans*. *J. Comp. Neurol.* **160,** 313–337.

Way, J. C., and Chalfie, M. (1988). *mec-3*, a homeobox-containing gene that specifies differentiation of the touch receptor neurons in *C. elegans*. *Cell* **54,** 5–16.

Way, J. C., and Chalfie, M. (1989). The *mec-3* gene of *Caenorhabditis elegans* requires its own product for maintained expression and is expressed in three neuronal cell types. *Genes Dev.* **3,** 1823–1833.

White, J., Southgate, E., Thomson, J., and Brenner, S. (1976). The structure of the ventral nerve cord of *C. elegans*. *Philos. Trans R Soc. Lond. B Biol. Sci.* **275,** 327–348.

White, J. G., Southgate, E., and Thomson, J. N. (1992). Mutations in the *Caenorhabditis elegans unc-4* gene alter the synaptic input to ventral cord motor neurons. *Nature* **355,** 838–841.

White, J. G., Southgate, E., Thomson, J. N., and Brenner, S. (1986). The structure of the nervous system of *Caenorhabditis elegans*. *Philos. Trans. R. Soc. Lond. B Biol. Sci.* **314,** 1–340.

Wrischnik, L., and Kenyon, C. (1997). The role of *lin-22*, a hairy/Enhancer of split homolog, in patterning the peripheral nervous system of *C. elegans*. *Development* **124,** 2875–2888.

Wu, J., Duggan, A., and Chalfie, M. (2001). Inhibition of touch cell fate by *egl-44* and *egl-46* in *C. elegans*. *Genes Dev.* **15,** 789–802.

Xue, D., Finney, M., Ruvkun, G., and Chalfie, M. (1992). Regulation of the *mec-3* gene by the *C. elegans* homeoproteins UNC-86 and MEC-3. *EMBO J.* **11,** 4969–4979.

Zhao, C., and Emmons, S. W. (1995). A transcription factor controlling development of peripheral sense organs in *C. elegans*. *Nature* **373,** 74–78.

SEXUAL BEHAVIOR OF THE *CAENORHABDITIS ELEGANS* MALE

Scott W. Emmons

Department of Molecular Genetics
Albert Einstein College of Medicine
Bronx, New York

Caenorhabditis elegans males express two sexual behaviors, mate-searching and copulation. Mate-searching is a spontaneous, goal-oriented behavior that models motivated behavior of higher animals. Copulation is the most complex behavior in *C. elegans*. It consists of a series of sub-behaviors flexibly linked together to achieve spicule insertion and sperm transfer. Sensory input associated with each sub-behavior has been identified. Specialized male-mating structures develop postembryonically from a largely common set of developmental genes. Significant exceptions are the DM domain transcription factors that act exclusively in males and contribute to the sexual dimorphism of the nervous system. Research into defining the molecular components of the neural circuits that underlie

INTERNATIONAL REVIEW OF
NEUROBIOLOGY, VOL. 69
DOI: 10.1016/S0074-7742(05)69004-6

0074-7742/06 $35.00

the copulatory sequence has been initiated. Studies of *C. elegans* male sexual behavior provide an opportunity to understand how a complex neural circuit functions to direct a purposeful series of actions directed towards a highly specific goal. They also provide an opportunity to study how distinct behavioral modes are integrated with one another within the context of the animal's overall physiological status and needs. Progress in these areas will require reconstruction of the male nervous system, an achievable goal with today's computer technology.

I. In *C. elegans*, Sex is the Male's Responsibility

C. elegans adult males express behaviors aimed at achieving fertilization of the opposite gender. While most behavioral studies carried out so far in *C. elegans* have focused on the hermaphrodite, taking advantage both of the facile hermaphrodite genetics and the availability of a complete wiring diagram of the hermaphrodite nervous system, male sexual behavior is increasingly a focus of attention. This is both because growing knowledge of the *C. elegans* nervous system and increasingly powerful genetic tools allow the complexity of sexual behavior to be addressed and because of the broad general interest in understanding sexual behavior. It is hoped that these studies will soon be augmented by a complete reconstruction of the male nervous system, efforts towards which have been initiated (S. W. Emmons, M. Xu, D. Albertson, N. Thomson, D. Hall, 2005, unpublished results, http://worms.aecom.yu.edu/).

In *C. elegans*, the two genders are a male and a self-fertile hermaphrodite. Hermaphrodites may have an ambivalent attitude towards sex. They can generate about 300 self-progeny genetically identical to themselves or they can mate and potentially generate a larger brood. They appear to express little or no behavior to promote sex, can produce only weakly attractive diffusible substances (Simon and Sternberg, 2002), and may even resist mating by darting away from males that are attempting to copulate with them, resisting spicule insertion, and expelling inseminated sperm (Barker, 1994; Kleemann, 2005; Liu and Sternberg, 1995). Male offspring (5AA, XO), which can appear spontaneously by X nondisjunction in the selfing hermaphrodite (5AA, XX), on the other hand, have an entirely different view of sex. They only pass on their genes if they mate and they express at least two behaviors that promote cross fertilization, mate-searching and copulation. Although males are rare in propagating populations, selective pressures must be such that functional copies of all genes necessary for male development and fertility have been maintained, although at least one natural population exists with mutated, infertile males (Chasnov and Chow, 2002; Hedgecock, 1976; Hodgkin and Doniach, 1997; Stewart and Phillips, 2002). In the laboratory, males placed on plates with hermaphrodites efficiently find them and copulate with them, their sperm preferentially fertilizing the hermaphrodite oocytes (Ward and Carrel, 1979).

II. Mate-Searching Demonstrates Motivated Behavior in the Worm

The *C. elegans* geneticist setting up a cross does not need to be concerned about where males are placed on the mating plate. Males will not remain in a region without hermaphrodites and will soon locate them. They appear to find mating partners by a random search behavior that is suppressed by a short-range hermaphrodite signals. Males therefore accumulate around hermaphrodites.

In order to search for hermaphrodites, males will overcome their natural attraction to food and will leave a food source if no hermaphrodites are present (Fig. 1A). Advantage has been taken of this fact to develop a quantitative "leaving" assay of mate-searching behavior in which the kinetics with which males leave a spot of bacteria is measured (Fig. 1B) (Lipton *et al.*, 2004). Mate-searching is regulated not only by the presence of hermaphrodites, but also by nutritional and reproductive status. Starved males will not leave food and both immature males and males lacking gonads or germ cells have reduced or absent mate-searching behavior (Lipton *et al.*, 2004). As a spontaneously expressed, goal-oriented behavior, *C. elegans* male mate-searching is an example of a motivated behavior (Kupfermann *et al.*, 2000). Studies of mutants lacking this behavior may lead to an understanding of how complex behavioral patterns of this sort are prioritized and regulated by environmental and physiological conditions (J. Lipton, R. Ghosh, G. Kleemann, L. Jia, C. Fang, S. W. Emmons, 2005, unpublished data).

III. Hermaphrodite Cues Influence Male Locomotory Behavior

In the leaving assay, if hermaphrodites are present mate-searching is suppressed and the male will associate with the hermaphrodites indefinitely. Thus, the male's behavior is influenced by the presence of hermaphrodites. This might be through direct physical contact or by a diffusible hermaphrodite signal. Such a signal must be short-lived because males begin to leave almost immediately (within 20 min) if the hermaphrodites are removed (Lipton *et al.*, 2004). The difference in behavior between a male on food with hermaphrodites and one without is not obvious. Inhibition of mate-searching is not simply a matter of

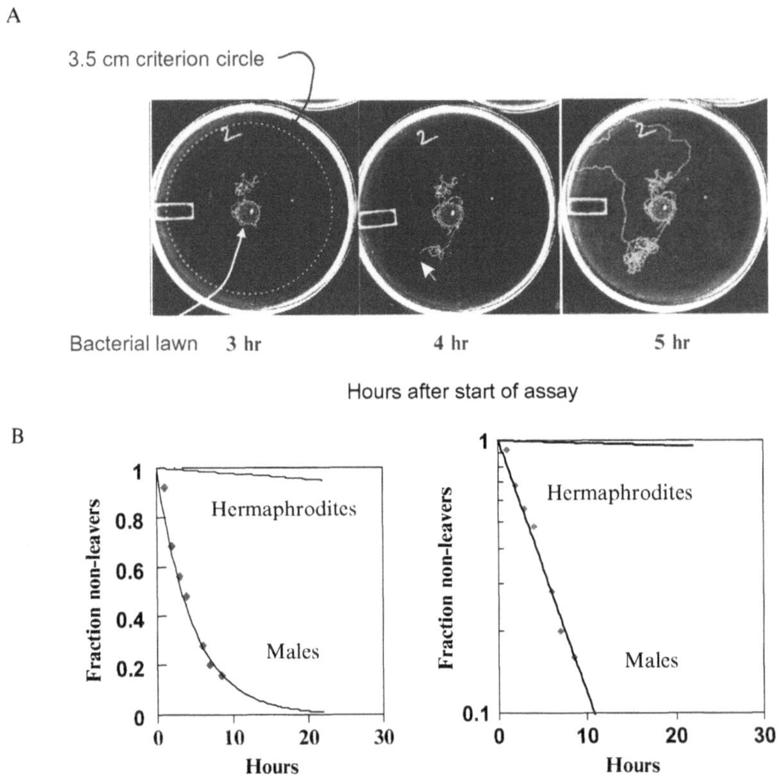

FIG. 1. Males leave food in a mate-searching assay. A. A plate in an assay for mate-searching ("leaving" assay) is shown at three times after setting up the assay. At time 0, a single male was placed on a 1 cm bacterial lawn on agar in the center of a 9 cm petri dish. As the male makes excursions off the lawn he leaves an indentation on the agar surface that can be photographed as a wiggly white line. During the first 3 h, the male made an excursion off the lawn and returned to it. At 4 h, the male was at the position indicated by the arrowhead. Between 4 and 5 h the male track traverses a 3.5 cm radius criterion circle and the male is scored as a leaver. B. Typical plots for 20 worms in a leaving assay of the fraction of worms that are non-leavers. Males (lower curves) leave following exponential kinetics (straight line on a semi log plot, right graph) while hermaphrodites remain on the food (upper curves). Part A is reproduced from Lipton *et al.*, 2004. Copyright 2004 by the Society of Neuroscience. (See Color Insert.)

constant copulation since for most of the time when hermaphrodites are present males are simply foraging on the lawn (Lipton *et al.*, 2004). In many nematode species, there is a strong pheromone attractant that brings the genders together (Green, 1980; Lee, 2002; Perry and Aumann, 1998) but in *C. elegans*, no such strong attractant can be detected. It is possible that the male detects a more general nematode signal (Coates and de Bono, 2002; de Bono and Bargmann, 1998; de Bono *et al.*, 2002), since other males also reduce the rate of leaving food.

In a second assay to demonstrate a hermaphrodite signal, male behavior is observed on agar preconditioned by hermaphrodites (Simon and Sternberg, 2002). In this assay, an increased rate of male reversals indicates the presence of a weak diffusible hermaphrodite signal. Whether this is the same signal that blocks leaving food in the mate-searching assay is not known.

IV. Male Copulatory Behavior can be Broken Down into a Series of Sub-Behaviors

C. elegans male copulatory behavior is thought to be the most complex *C. elegans* behavior. Copulation is triggered when a male contacts a hermaphrodite with his tail (Fig. 2). In response to such contact, the male halts his forward locomotion and

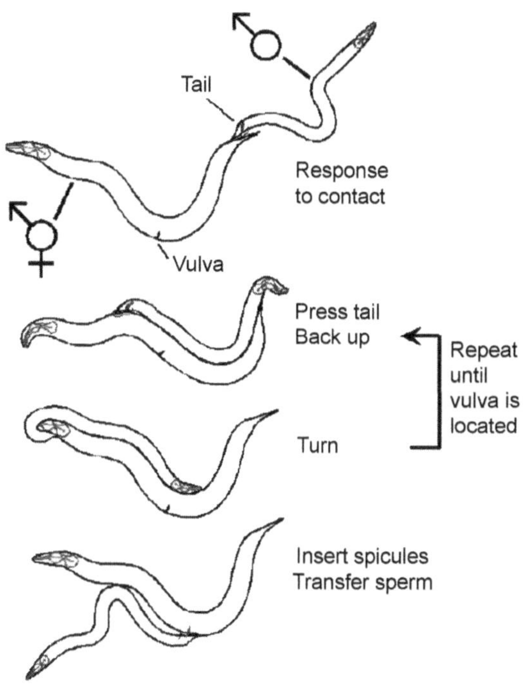

FIG. 2. Male copulatory behavior. The *C. elegans* male expresses a sequence of stereotyped behaviors in order to locate the hermaphrodite vulva, insert his spicules, and transfer sperm.

presses his tail against the hermaphrodite body. He then backs up, scanning the body for the vulva. If in backing he reaches either end, he executes a sharp ventral bend while maintaining contact and scans the other side. When the vulva is detected, backing stops and the vulva is probed by rapid prodding with hardened spicules. During this period, the male synchronizes his forward and backward locomotion to the hermaphrodite's in order to remain in contact with the vulva. If the spicules breach the vulval opening, they are inserted into it and remain there while ejaculation occurs, which requires a few seconds. After approximately 1 min, the spicules are withdrawn and the male swims away. He remains in a refractory state lasting about 20 min before he will copulate again.

This multi-step copulatory sequence can be broken down into a series of sub-behaviors (Barr and Sternberg, 1999; Emmons and Sternberg, 1997; Garcia *et al.*, 2001; Liu and Sternberg, 1995; Loer and Kenyon, 1993) (Fig. 3). Existence of sub-behaviors is supported by both cell ablation and genetic studies as described in the following text. These copulatory sub-behaviors are viewed in the following way. Each sub-behavior consists of a discrete, self-reinforcing behavioral state of the nervous system. A behavioral state is associated with and reinforced by a particular condition of sensory inputs that is reached at a point on the pathway towards successful ejaculation into the uterus. A sub-behavior continues for as long as this sensory input remains within a certain range. But the sub-behavior tends to move the male to a position where a new set of inputs will be received. When this happens, a switch in behavioral state is triggered,the ongoing behavioral state is inhibited, and a new behavior is initiated. This process repeats as it takes the male through the successive steps of the copulatory sequence.

These copulatory sub-behaviors need not occur in a fixed sequence or even occur at all; a copulating male is remarkably persistent and adaptable in his efforts to find the vulva. For example, contact with the vulva supersedes all the earlier steps and immediately jumps the sequence ahead to the prodding phase. If contact with the vulva but not the hermaphrodite body is lost, the program backs up to the earlier scanning phase of backwards locomotion. If contact is lost altogether, the male returns to forward locomotion. Turning occurs only if the end of the hermaphrodite is detected.

The view of copulation as consisting of a small number of discrete sub-behaviors is a simplification. The sub-behavioral states themselves may be complex and have within them their own sub-sub-behaviors. For example, when probing for the vulva, the male moves backwards and forwards to follow the hermaphrodite's movements. Thus, this sub-behavior contains within it further subdivisions of forward and backward locomotory states. Understanding how a set of stable behavioral states are flexibly linked together and coordinated by sensory inputs into a purposeful behavioral series is one of the interesting challenges in the study of *C. elegans* male copulatory behavior.

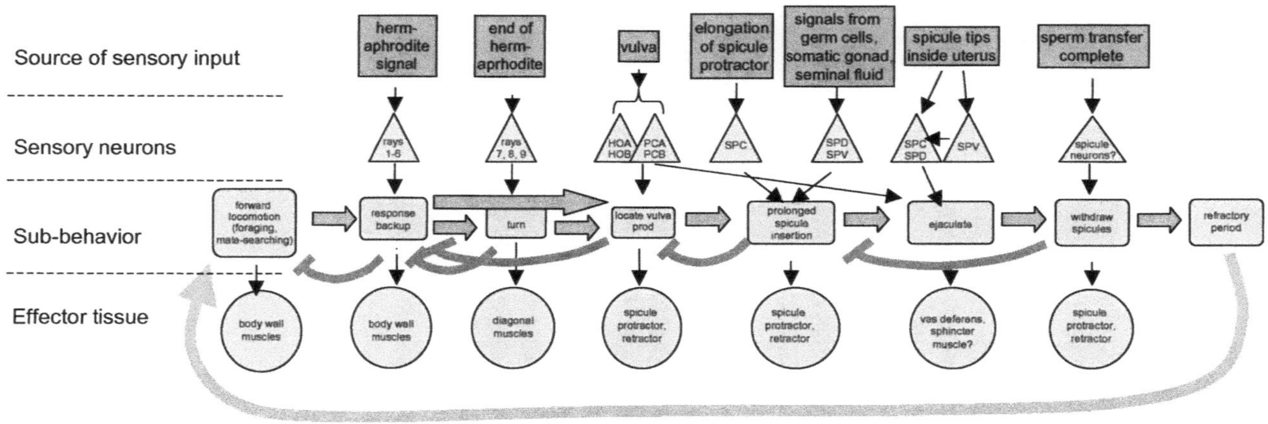

FIG. 3. Copulation proceeds through a series of sub-behaviors. Each sub-behavior is initiated and reinforced by sensory input through identified sensory neurons. Although the sources of signals are shown, none of the molecular or mechanical cues detected by sensory neurons have been identified to date. The most prominent sensory input associated with each sub-behavior is shown; some sub-behaviors can be sustained by alternative input. For example, the hook (HOA, HOB) and post-cloacal sensilla (PCA, PCB, PCC) can each sustain the response and back up sub-behaviors. Each sub-behavior tends to lead the male to a new set of inputs, which causes a switch in behavior, the ongoing sub-behavior is suppressed, and a new sub-behavior is initiated. The pathway need not occur in fixed order. It can jump over steps or back up in the sequence as necessary. (See Color Insert.)

V. The Gender-Specific Anatomy and Nervous System of the Male Develop Postembryonically

Male copulatory behavior is subserved by a set of anatomical structures, special muscles, and neural circuitry. These elements develop postembryonically (Sulston and Horvitz, 1977; Sulston *et al.*, 1980). Larval males resemble larval hermaphrodites and although the larval nervous system has not been explicitly reconstructed, it is probably not sexually specialized because the behavioral requirements for the larval stages are not gender-specific. Larval males do not mate-search and do not respond to hermaphrodites.

Four male-specific sensory neurons in the head, the CEM neurons, which are born during embryogenesis, appear to undergo a maturation step at the end of larval development (White *et al.*, 1986). The remaining 85 male-specific neurons arise from postembryonic cell lineages and mostly come to reside in a series of ganglia in the male tail that are much enlarged over those in the hermaphrodite (Fig. 4A). The male sexual neurons are more highly branched than most of the non–sex-specific neurons and construct circuitry, mostly in the pre-anal ganglion, comparable in complexity to that found in the anterior nerve ring (Sulston *et al.*, 1980). In addition, postembryonic myogenic lineages are sexually dimorphic and result in the generation of 41 male-specific sex muscles (Fig. 4B). At the final larval-to-adult molt, the morphology of the tail is drastically reorganized by retraction of the terminal cells, generating the scalloped shape of the tail, the acellular fan, and the four classes of sensory structures: the rays, the spicules, the hook, and the post-cloacal sensilla (Nguyen *et al.*, 1999; Sulston *et al.*, 1980) (Fig. 4C). In addition to these male-specific cells, further changes occur that affect the cells found in both genders. Some common neurons are more highly branched and connected in the male, and the sphincter muscle reorganizes (Sulston *et al.*, 1980). Some male-specific neurons grow out of the tail and send an axonal process through the ventral nerve cord to the head and into the nerve ring. The nature of the connections these neurons make and the degree of gender-specific new wiring among the common set of anterior neurons remains to be determined.

VI. The Fan Facilitates the Male's Contact with the Hermaphrodite Body

Each anatomical specialization of the adult male tail functions to facilitate one or more of the steps of the copulatory behavioral sequence. Most obvious among these structures are the fan and rays. Phylogenetic analyses show a significant correlation between parallel mating position and presence of a broad

A

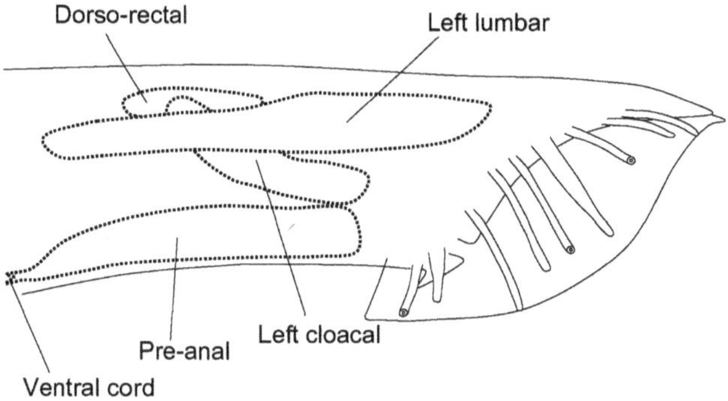

B

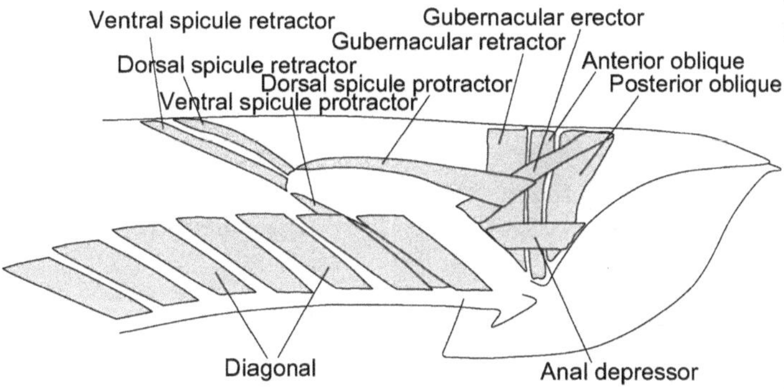

C

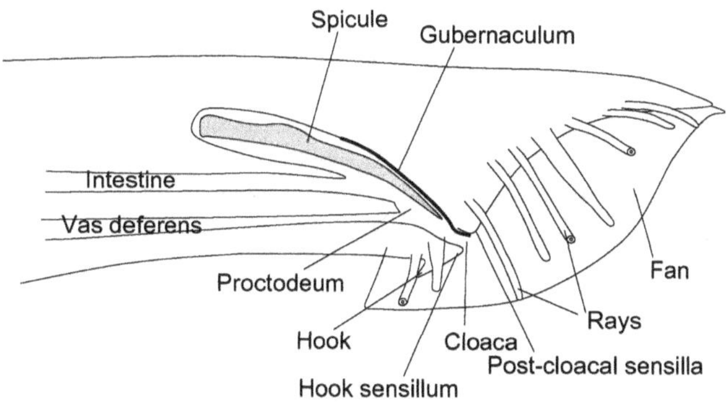

FIG. 4. The major specializations of the adult male tail. A. Ganglia. Lumbar and cloacal ganglia are in left-right pairs. B. Male-specific sex muscles. In addition to those muscles shown, 10 male-specific longitudinal muscles take up positions next to the existing ventral non–sex-specific body wall muscles. C. Sensilla. The four major classes of sensilla are the 9 pairs of rays (neurons RnA, RnB), the hook sensillum (neurons HOA, HOB), the left and right post-cloacal sensilla (neurons PCA, PCB, PCC), and the left and right spicules (neurons SPC, SPD, SPV). The fan is an acellular webbing of cuticle. The proctodeum is the chamber where the gut and gonad join at the cloaca. The gubernaculum is a sclerotic layer of cuticle that presumably functions to guide the spicules out through the cloaca.

fan (Fitch, 2000). In "parallel" mating, mating partners lie side by side, whereas in "spiral" mating, the male curls his tail circumferentially around the female body (Sudhaus and Fitch, 2001). Parallel mating may correlate with worms living on the surface of a solid substrate, where both animals are constrained to lie on their sides (because flexure of the body is constrained to the dorsoventral plane). The mid-ventral vulva is directed laterally and the male need only to perform a one-dimensional search around the hermaphrodite dorsoventral midline to locate it. The fan serves to keep the male oriented with his ventral side facing the hermaphrodite during this search. It may also help convey hermaphrodite contact, through the rays, by being forced open from a collapsed condition it would tend to take when the male is isolated. By connecting the rays together and constraining them to move as one, the fan may result in a robust open versus closed signal. This model assumes that at least one of the two ray sensory neurons is mechanosensory, an assumption that has not been demonstrated but is supported by the observation that males will attempt to copulate with physical objects that would not be expected to have chemosensory cues, such as fibers or the agar surface. Other nematodes that, unlike *C. elegans*, can mate in liquid generally lack fans and mate in the "spiral" position (Fitch, 2000). To find the vulva, they must search the entire two-dimensional hermaphrodite surface.

VII. Each Sub-Behavior of the Copulatory Sequence is Promoted by Sensory Input from Particular Sensory Neurons

The four sets of male-specific sensilla in the male tail are situated appropriately within the mating structures to monitor and promote the steps of the copulatory sequence. The functions of individual sensory neurons have been explored by means of cell ablation studies utilizing a laser microbeam (Garcia *et al.*, 2001; Liu and Sternberg, 1995). The results provide experimental support for a model consisting of sub-behaviors (Fig. 3). Roughly, the rays function in the initial response and search for the vulva, the hook sensillum and post-cloacal sensilla detect the vulva, and the spicule neurons monitor the disposition of the spicules and control ejaculation. However, some sub-behaviors require an appropriate combination of inputs from multiple sensilla, while others may be sustained by alternative sets of inputs. Reconstructions of the male circuitry reveal a substantial degree of cross-wiring among the sensory neurons that would support interactions among the various sensory inputs (Sulston *et al.*, 1980; S. W. Emmons, M. Xu, D. Albertson, N. Thomson, D. Hall, 2005, unpublished results). The behavioral analysis reveals a high degree of redundancy and flexibility that is no doubt necessary for the achievement of this robust behavior.

VIII. The Rays Convey Complex Spatial Information

The primary roles of the rays are in the response and backing steps (Fig. 2), in which the male halts forward locomotion, places his tail in a distinctive posture against the hermaphrodite, and begins to back up (Fig. 5A). This function of the rays is reinforced by the other ventrally located sensilla, the hook, postcloacal, and spicule sensilla, since males lacking rays can still respond to hermaphrodite contact and mate, albeit poorly, via these alternate inputs (Liu and Sternberg, 1995). Each ray contains two ciliated sensory neurons with different ultrastructure, denoted A type (RnA) and B type (RnB); both types probably respond to a variety of chemical and mechanical cues. In most rays, the B-type neurons have their sensory ending

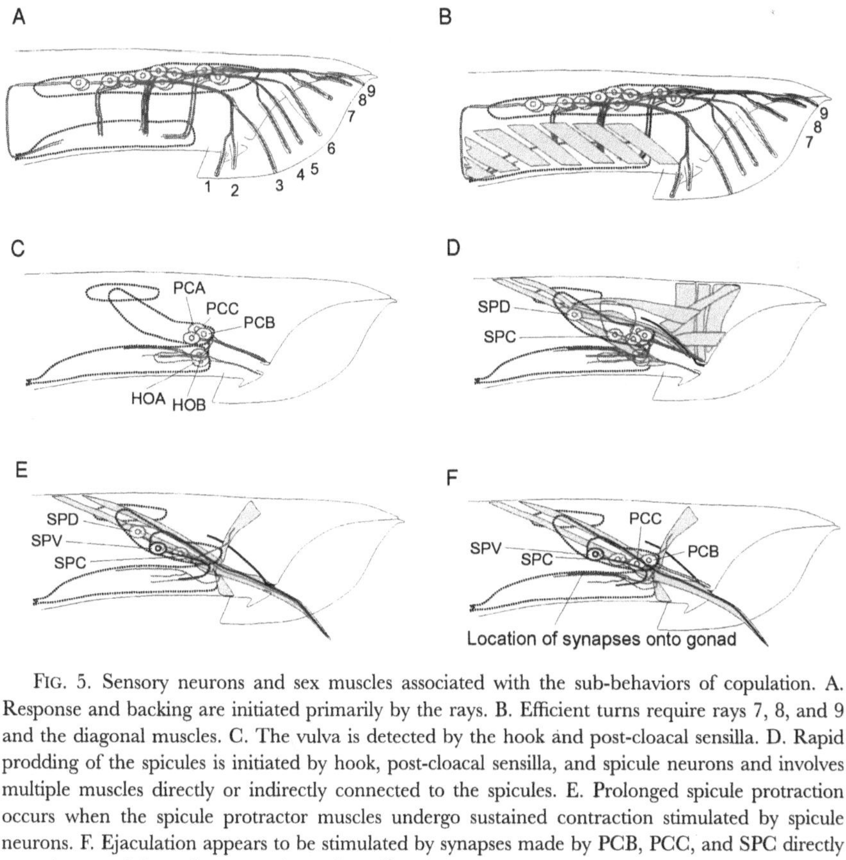

FIG. 5. Sensory neurons and sex muscles associated with the sub-behaviors of copulation. A. Response and backing are initiated primarily by the rays. B. Efficient turns require rays 7, 8, and 9 and the diagonal muscles. C. The vulva is detected by the hook and post-cloacal sensilla. D. Rapid prodding of the spicules is initiated by hook, post-cloacal sensilla, and spicule neurons and involves multiple muscles directly or indirectly connected to the spicules. E. Prolonged spicule protraction occurs when the spicule protractor muscles undergo sustained contraction stimulated by spicule neurons. F. Ejaculation appears to be stimulated by synapses made by PCB, PCC, and SPC directly onto the gonad from the pre-anal ganglion. (See Color Insert.)

situated at an opening to the environment at the ray tip (Sulston *et al.*, 1980). One G-protein–coupled, 7 transmembrane receptor (*srw-140*) has been identified that appears to be expressed in the B type neurons (Portman and Emmons, 2004). Since the male responds when his fan touches a hermaphrodite, the ligand for such a receptor may be one or more moieties on the hermaphrodite cuticle. Hermaphrodite mutants with abnormal cuticle antigens have been identified that evoke a poor response from wild type males (J. Wang, M. M. Barr, 2005, personal communication). The B type neurons express TRP family channels (LOV-1, PKD-2) that are thought to be mechanosensory (see the following text) (Barr and Sternberg, 1999). This suggests that the protruding tips of the B-type neurons might respond to mechanical touch. The dendritic endings of the A-type neurons are located somewhat inside the ray tip exposed to the exterior through the channel. Ray 6 has a thicker overall shape than the other rays, no opening or only a tiny one, and a B-type neuron with slightly different ultrastructure (Sulston *et al.*, 1980). The significance of ray 6 specialization is unknown.

The fan and rays together may be considered to be a compound sensory organ that conveys topographic information about the location of various sensory inputs to central circuitry. Possibly the clearest evidence for this complex function comes from analysis of the function of individual rays in the next step of the copulatory sequence, the turn around the end. For execution of this sub-behavior, the most posterior rays, rays 7, 8, and 9, are specifically required (Liu and Sternberg, 1995) (Fig. 5B). It seems likely that the male senses the end of a hermaphrodite when he detects differential inputs between these posterior rays and more anterior rays. The male often turns just before he actually reaches the end, when the body starts to taper. Rays 7, 8, and 9 must either sense molecular moieties in the terminal region of the cuticle or mechanically sense the taper as the posterior fan region starts to close.

The turn itself requires a set of six male-specific motor neurons, the CP neurons. These innervate the diagonal muscles and are serotonergic. Exogenous serotonin induces a tail-curling reflex similar to that seen during mating. Loer and Kenyon (1993) confirmed the role of diagonal muscles, CP neurons, and serotonin in turn beahavior in a study employing cell ablations as well as analysis of serotonin-deficient mutants.

The rays not only sense differential inputs along the anteroposterior axis, but also between the dorsal and ventral surfaces of the fan. The openings of the rays are placed at reproducible, genetically determined positions, with the tips of rays 1, 5, and 7 opening on the dorsal surface, rays 2, 4, and 8 opening on the ventral surface, and rays 3 and 9 opening right at the fan margin (Baird *et al.*, 1991; Sulston *et al.*, 1980). The dorsally opening rays mediate the male's response to dorsal touch (Liu and Sternberg, 1995).

Further support for differential functions of the rays comes from evidence of their molecular complexity. Serotonin, dopamine, and FMFRamide

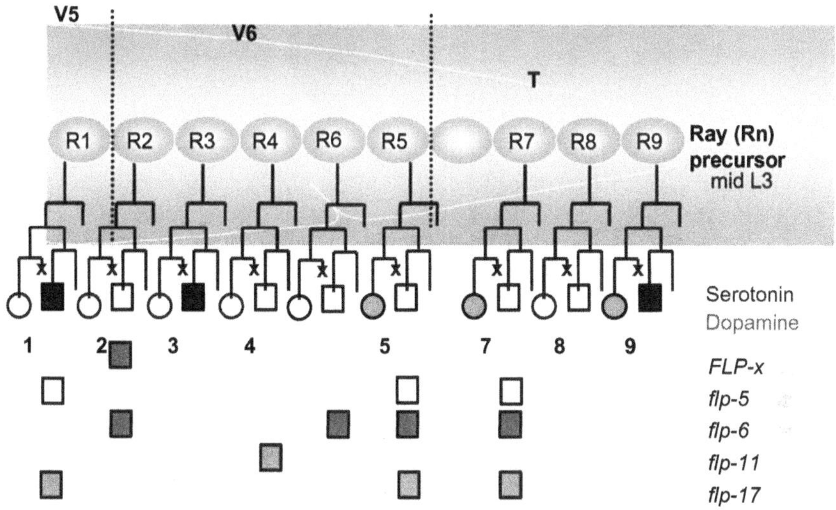

FIG. 6. Neurotransmitters expressed by the rays. Ray precursor cells (Rn) in the posterior male seam express the ray sublineage during the late L3 early L4 larval stages as shown. Rn.aaa: A-type neuron (circle); Rn.aap: programmed cell death; Rn.apa: B-type neuron (square); Rn.app: structural cell of the ray; Rn.p: hypodermal cell. The *flp* genes encode FMRFamide peptide neurotransmitters. Adapted from Lints *et al.* (2004). (See Color Insert.)

neurotransmitters are expressed in specific subsets of the rays (Lints and Emmons, 1999; Lints *et al.*, 2004) (Fig. 6). Males deficient in either serotonin or dopamine mate poorly; in particular, they execute sloppy turns (Liu and Sternberg, 1995; Loer and Kenyon, 1993). The 7-transmembrane receptor *srd-1* is expressed in ray 9 (Troemel *et al.*, 1995). These differences in expression patterns of neuronal genes among the rays are brought about by differential expression of a set of transcription factors in a pattern that is established during the postembryonic cell lineages (Chow and Emmons, 1994; Chow *et al.*, 1995; Lints and Emmons, 1999; Lints *et al.*, 2004; Salser and Kenyon, 1996; Zhang and Emmons, 1995). In addition, a spatially restricted transforming growth factor-beta (TGFβ) signal affects development of rays 5, 7, and 9 (Lints and Emmons, 1999) (Fig. 7). The transcription factors and TGFβ signal serve to coordinately determine the specific properties of all three cells of the rays and hence may be viewed as ray identity genes (Lints *et al.*, 2004).

As might be expected from this functional and molecular complexity, initial studies of neuronal connectivity in the male tail indicate that the axons of neurons from different rays project to different sets of synaptic targets (S. W. Emmons, M. Xu, D. Albertson, N. Thomson, D. Hall, 2005, unpublished results). The unknown molecular functions that specify these differential targets are presumably

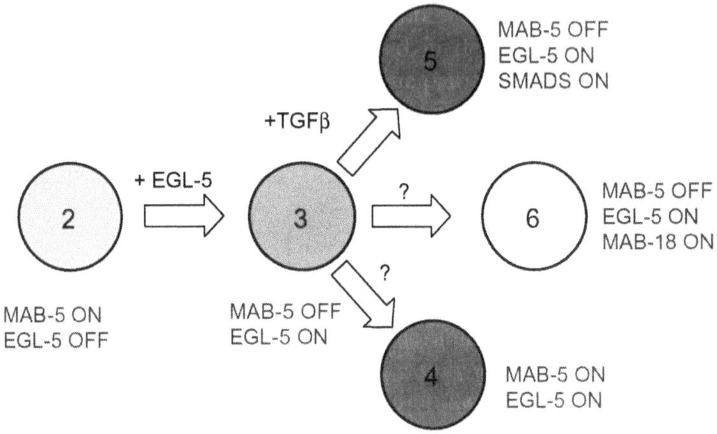

FIG. 7. A transcription factor code establishes the identities of the V6 rays (the rays descended from the L1 blast cell V6). This schematic diagram shows the hierarchical relationships among the identities of the V6 rays. The differing characteristics of the V6 rays, rays 2 to 6, are determined in part by the Hox genes *mab-5* and *egl-5*, the Pax6 gene *mab-18*, and a TGFβ signal. In *egl-5* mutants, ray 6 is not made because of a lineage transformation while the remaining rays assume the characteristics of ray 2, which may be considered a ground state. The TGFβ signal is selectively received by ray 5. Further signals involved in establishing the identities of rays 4 and 6 remain to be identified. Reprinted from Lints *et al.* (2004), with permission from Elsevier. (See Color Insert.)

also controlled by the ray identity transcription factors. Each ray sensory neuron makes multiple chemical and electrical synapses in several branches to partners in both the lumbar ganglia and the preanal ganglion. Partners include other ray neurons, sensory neurons of other sensilla, interneurons, and even muscles, reflecting the highly interactive and complex nature of the circuitry.

IX. Hook, Post-Cloacal Sensilla, and Spicules Detect the Vulva

Contact with the vulva ends the backing sub-behavior and initiates a new sub-behavior in which the male attempts to insert his spicules into the vulva. The male stops backing and the spicules rapidly prod (7 Hz) the hermaphrodite body. If they penetrate the vulval opening, prodding stops and the next sub-behavior initiates, a prolonged spicule protraction into the uterus accompanied by ejaculation (Fig. 3) (Garcia *et al.*, 2001). The vulva is detected by both the hook sensillum and the paired post-cloacal sensilla via unknown chemical and possibly mechanical cues (Liu and Sternberg, 1995) (Fig. 5C). Thus, these same neurons can act both to promote one sub-behavior—backing—and to terminate this behavior and initiate a different sub-behavior—prodding. In addition to hook and post-cloacal neurons, the prodding phase is promoted by spicule neurons SPD and SPC (Fig. 5D).

During the prodding phase, the male maintains contact with the vulva by swimming backwards and forwards coordinately with the hermaphrodite's movements. Timing of the male's reversals may be triggered by comparison of inputs from hook and post-cloacal sensilla; in order to keep both sets in contact with a vulval signal, the male swims forward if only the hook detects the signal and backwards if only the post-cloacal sensilla detects the signal (R. Garcia, 2005, personal communication). Comparison with input from ray neurons might also be involved, since there are cross-connections between both hook and post-cloacal sensillum neurons with specific subsets of ray neurons (S. W. Emmons, M. Xu, D. Albertson, N. Thomson, D. Hall, 2005, unpublished results). One possible function of the hook itself, which is a hardened point of cuticle, might be to scrape off a copulatory plug previously deposited by an earlier mating male (see the following text) (R. Garcia, 2005, personal communication).

X. Spicule Insertion Anchors the Male and Triggers Ejaculation

Successful spicule insertion appears to be detected by two cues: contraction of the spicule protractor muscle and substances inside the uterus. Contraction of the spicule protractor muscle is detected by the SPC neuron, which has an apparently proprioceptive dendritic ending associated with this muscle (Sulston *et al.*, 1980). Chemical cues of the uterus can be detected by the sensory dendrites of SPD and SPV, both of which are situated at an opening in the end of each spicule. Ejaculation occurs when the spicules are inserted and the post-cloacal sensilla detects the vulva (Fig. 5F). The function of SPV is to prevent SPD and SPC from triggering ejaculation prematurely. In the absence of SPV, ejaculation may occur during the prodding phase, dependent upon the function of SPC and SPD (Liu and Sternberg, 1995).

Ejaculation may be triggered by direct signaling between sensory neurons and the vas deferens (Fig. 5F). Post-cloacal sensilla neurons PCB and PCC, and spicule neurons SPC (SPD and SPV have not been reconstructed) make synapses against the dorsal basement membrane of the pre-anal ganglion, which lies directly underneath the vas deferens (S. W. Emmons, M. Xu, D. Albertson, N. Thomson, D. Hall, 2005, unpublished results). Release of neurotransmitter across this basement membrane, thus, may induce ejaculation. The mechanism of ejaculation itself is unknown. There are no muscles positioned in such a way as to provide an ejaculatory force (Sulston *et al.*, 1980). Along the wall of the vas deferens, clear bubble-like structures are visible with Nomarski microscopy that might contain and release the ejaculate.

Males of many species and some strains of *C. elegans* deposit an adherent, gelatinous plug at the hermaphrodite vulva after they retract their spicules but before they swim away (Barker, 1994; Hodgkin and Doniach, 1997). The commonly used laboratory strain of *C. elegans* (N2) may be mutant for this function as the plugging allele of the locus *plg-1* is dominant (Hodgkin and Doniach, 1997). *plg-1(+)* may cause coagulation of a substance in seminal fluid and function either to hinder subsequent maters or to prevent sperm expulsion and promote male fertility (Barker, 1994; Barr and Garcia, 2005). However, attempts to demonstrate such effects were not successful (Hodgkin and Doniach, 1997).

XI. The Male's Copulatory Nervous System can Take Over Control of the Rest of the Nervous System

Throughout the mating sequence, the entire male nervous system must be brought under the control of signals from the male-specific sensory neurons in the tail in order to prevent simultaneous expression of alternative behavioral modes that would interfere with copulation. For example, when the rays detect a hermaphrodite signal, forward locomotion is stopped and behavioral modes associated with foraging or mate-searching must be suppressed while the male is engaged in copulating. During copulation, pumping of the pharynx continues during the backing phase but ceases when the male's spicules are inserted (R. Garcia, 2005, personal communication). After copulation, the function of the copulatory apparatus itself is apparently suppressed since the male will not mate during a short refractory period. Conversely, mating is regulated by the remainder of the nervous system or by a physiological state since males that have been without food for some time, although they remain healthy and vigorous and contain sperm, will not mate. These observations indicate cross-communication between different circuits within the nervous system. Such cross-communication is no doubt mediated by humoral signals as well as by direct interconnections. Extensive synaptic interactions between male-specific neurons and the common set of neurons occur in the tail, particularly in the pre-anal ganglion. In addition, an unknown degree of cross-connectivity is made by several male-specific interneurons with cell bodies in the tail which send processes through the ventral nerve cord into the nerve ring (Sulston *et al.*, 1980).

XII. Male Development Results from Gender-Specific Utilization of a Largely Common Set of Genes

Genes underlying male development and behavior have been sought using genetic and molecular means. It appears from the results of these studies that the extensive differences between male and hermaphrodite development and behavior do not arise from a large number of gender-specific genes. Rather, male development and behavior by and large uses the same genes as hermaphrodite development and behavior, employed in male-specific developmental programs or neuronal circuits. In whole-genome microarray surveys comparing messenger ribonucleic acid extracted from males and hermaphrodites, only 759 genes (4%) were enriched above two-fold in the male soma (Jiang *et al.*, 2001). Of these, 14 were transcription factors that might play a more significant role in male development, while 40 represented genes with known neuronal functions and might be important in male behavior. Among a collection of ray-expressed genes identified in microarray experiments, only the *cwp* genes were male-specific (see the following text) (Portman and Emmons, 2004).

A similar result has been obtained by genetic analysis. The small number of published screens for male mutants have mostly uncovered genes that function in both genders (Baird and Emmons, 1990; Baird *et al.*, 1991; Barr and Sternberg, 1999; Chow and Emmons, 1994; Garcia and Sternberg, 2003; Hodgkin, 1983). Examples of major non–sex-specific regulatory or differentiation genes identified in screens for abnormal males (*mab* genes for *male abnormal*) include the Hox gene *mab-5* (Hodgkin, 1983), semaphorin (*mab-20*) (Baird *et al.*, 1991), and ephrin (*efn-4/mab-26*) (Chow and Emmons, 1994). Significant exceptions are the male-specific polycystin genes *lov-1* and *pkd-2*, discussed further in the following text (Barr and Sternberg, 1999). Conversely, many genes with significant effects on male development or behavior have been identified by studying the male phenotypes of mutations first identified by their effects in hermaphrodites. For example, the Hox gene *egl-5*, which is essential for patterning the various structures of the male tail, was first identified because of its role in hermaphrodite egg-laying. The only significant, apparently completely male-specific regulatory transcription factors known are the DM domain transcription factors *mab-3* and *mab-23*.

XIII. DM Domain Transcription Factors may Represent Primordial Male-Determining Genes

Two genes identified in screens of male abnormal mutants in *C. elegans*—*mab-3* and *mab-23*—are members of a conserved family of sexual regulatory transcription factors. In *C. elegans*, gender is determined by sex-chromosome dosage, which functions to regulate the level of activity of the Zn finger master sex regulatory transcription factor TRA-1 (Meyer, 1997). TRA-1 and the upstream pathway that regulates it is not found in the gender-determination systems of other groups of animals. MAB-3 and MAB-23, on the other hand, are members of a conserved family of Zn-binding transcription factors, so-called DM-domain transcription factors named for the founding members *mab-3* and *Drosophila doublesex* (Raymond *et al.*, 1998). Proteins of the DM domain family function in defining aspects of the male phenotype in worms, flies, fish, and mammals (Matsuda *et al.*, 2002; Raymond *et al.*, 1998, 2000). Mutations in both *mab-3* and *mab-23* were identified by their effects on the rays (Lints and Emmons, 2002; Shen and Hodgkin, 1988). Both are also required for the differentiation of multiple male tissues and efficient mating behavior but appear to have no discernable function in hermaphrodites. Indeed, an allele of *mab-23* causing male infertility due to multiple abnormalities in male tail structures was found initially in a natural population, attesting to its lack of significant deleterious effect on the hermaphrodite (Hodgkin and Doniach, 1997).

Hodgkin (1992) has pointed out that modern gender-determination systems may have evolved from primordial systems consisting of a dominant male-determining factor. Conserved function in male differentiation since the time of an ancient ancestor together with absence of any apparent function in hermaphrodites in *C. elegans* (although there is a function of *doublesex* in female differentiation in *Drosophila*) suggests that MAB-3 and MAB-23 may represent members of such a primordial male-determining family. Perhaps counterintuitively, these genes do not appear to regulate whole programs of male development, but rather act at the end of such programs to facilitate specific aspects of differentiation. For example, *mab-23* has in part the subtle effect of modulating the induction of dopamine and serotonin by a TGFβ signal in certain ray neurons (Lints and Emmons, 2002). Where they act in a tissue that is present in both genders, DM domain genes may be directly regulated by TRA-1. For example, *mab-3* expression, which blocks yolk gene transcription in the male gut, is directly repressed by TRA-1 in the hermaphrodite gut (Yi *et al.*, 2000). However, in tissues with extensive male-specific differentiation, expression of *mab-3* and *mab-23* occurs in the hermaphrodite as well as in the male but has an effect only in the male (Lints and Emmons, 2002; Yi *et al.*, 2000). In such tissues, the regulatory effects of TRA-1 must come at a higher level in the developmental hierarchy and target non–sex-specific regulatory genes. Since male-specific tissue is not made, the DM domain genes become irrelevant. Multiple defects in male-mating behavior, particularly in *mab-23* mutants, suggests that DM domain transcription factors may play important roles in the overall gender-specific differentiation of the nervous system (Lints and Emmons, 2002).

XIV. Response to Contact and Vulva Location Require Genes of the TRP Channel Superfamily of Putative Mechanosensitive Channels

To identify genes involved in the function of the neuronal circuits that underlie mating behavior, analysis of existing mutants as well as screens for new mutants can be undertaken for males defective in specific sub-behaviors of the copulatory sequence. Analysis of mutants defective in serotonin biosynthesis revealed a role for this neurotransmitter in the turn sub-behavior, as previously mentioned (Loer and Kenyon, 1993). Analysis of cilium-defective mutants verified a role of ciliated sensory neurons in response and vulva location behavior (Barr and Sternberg, 1999). Isolation of new mutants defective in response and vulva location behavior resulted in the identification of two genes, *lov-1* and *pkd-2,* related to polycystic kidney disease genes and now known to be members of the TRP superfamily of Ca^{++}-permeable channels (Barr and Sternberg, 1999; Corey, 2003). The possible reason for this at first surprising connection is revealed by studies of channel function. Located in cilia, human PKD1 and PKD2, the counterparts of worm LOV-1 and PKD-2, respectively, sense bending of the cilium as fluid passes over the cell surface (Nauli *et al.*, 2003). LOV-1 and PKD-2 are similarly localized to the ciliated endings and cell bodies of the B-type neurons of the rays, in HOB, and in the male-specific CEM sensory neurons of the head (Barr and Sternberg, 1999; Barr *et al.*, 2001). Localization in the rays and hook sensillum is consistent with the assigned functions of these sensilla in the response and vulva location steps of copulation and with a possible mechanosensory role for these structures. Recent microarray studies have identified five novel genes, *cwp-1* through *cwp-5*, that are coexpressed with the polycystins in the RnB, HOB, and CEM neurons (Portman and Emmons, 2004). These genes may have regulatory roles in the function of this mechanosensory signaling pathway.

XV. Multiple Cholinergic Pathways Regulate the Muscle Contractions Controlling Spicule Behavior

Control of the motor circuits governing the spicule movements has been dissected by a combined approach using existing and new mutants, pharmacological studies, and cell ablations (Garcia *et al.*, 2001). Neurotransmitter pathways involved in spicule function were identified by analyzing existing mutants in cholinergic neurotransmission (Fig. 8). The membrane-localized L-type voltage-gated calcium channel EGL-19 is required for the prolonged contraction of the spicule protractor muscle that sustains spicule insertion during ejaculation.

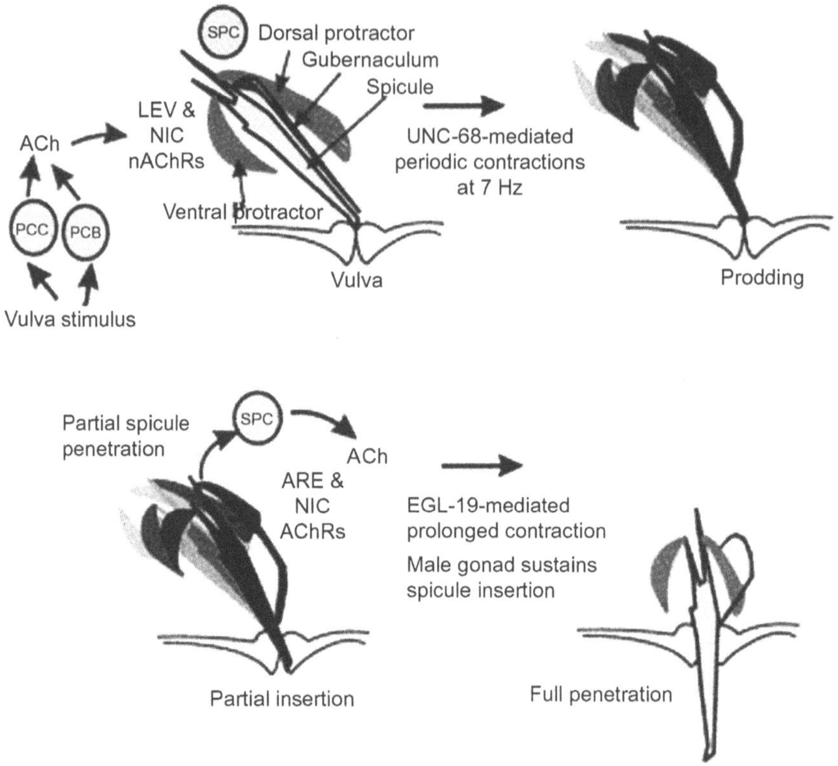

FIG. 8. The roles of two cholinergic pathways in prodding and penetration behavior of the spicules. Top. Stimulated by ACh from post-cloacal sensilla neurons PCC and PCB and spicule neuron SPC, the spicule protractor muscle twitches, causing the spicules to rapidly prod the region of the vulva. Prodding involves release of intracellular Ca^{++} by the ryanodine-sensitive, sarcoplasmic Ca^{++} channel UNC-68. Bottom. When the spicules penetrate the vulva, SPC, which has a presumptively proprioceptive dendrite attached to the spicule protractor muscle, releases ACh and stimulates a prolonged contraction. Prolonged protraction involves entry of extracellular Ca^{++} through the L-type voltage-gated channel EGL-19 and is sustained by substances in seminal fluid. Reprinted from Garcia *et al.* (2001), with permission from Elsevier. (See Color Insert.)

This channel functions downstream of an unidentified receptor sensitive to the acetylcholine agonist arecoline. It acts in muscle, likely the protractors. Prolonged contraction of the protractors is stimulated by endogenous acetylcholine. One source of acetylcholine is the SPC sensory motor neuron, which has both a presumptively proprioceptive dendrite attached to the muscle and motor output onto it (Sulston *et al.*, 1980). Presumably, once the spicule protractor muscle shortens beyond a certain point because the spicules have succeeded in penetrating the vulva, Ca^{++} entry from the extracellular environment through EGL-19 channels triggers a prolonged and sustained contraction. The time of such contraction is further lengthened by substances in seminal fluid and sperm, presumably detected as the male ejaculates by one or both of the chemosensory spicule neurons SPD and SPV.

A different source of Ca^{++} is involved in the periodic contraction sub-behavior that precedes prolonged insertion. This source releases intracellular Ca^{++} through the sarcoplasmic, ryanodine-sensitive channel UNC-68. UNC-68 functions downstream of the nicotinic acetylcholine receptor UNC-38/UNC-29, which can be activated by the acetylcholine agonist levamisole. UNC-68 function is also in muscle, but which muscle is uncertain. Periodic contraction of the spicule protractor muscle can be stimulated by sensory input from the hook neurons and post-cloacal sensilla neurons. Among these, PCB and PCC are cholinergic, but PCB and PCC do not appear to synapse onto the spicule protractors (S. W. Emmons, M. Xu, D. Albertson, N. Thomson, D. Hall, 2005, unpublished results). They do have synapses onto the SPC neurons, but SPC neurons are not required for prodding behavior. One possibility is that PCB and PCC stimulate the spicule protractors via the oblique muscles (Fig. 4B), which do receive synaptic input from PCB and PCC and which are observed to twitch along with the protractors during prodding behavior (Barr and Garcia, 2005). Oblique muscles might have gap junctions onto the spicule protractors in the way that vulval muscles are joined together by gap junctions. Further reconstruction of the male tail will reveal whether this speculation is correct.

Many additional molecular components are undoubtedly involved in the regulation of the spicule circuitry. One protein necessary to prevent triggering of the circuit inappropriately is an ERG-like potassium channel encoded by the gene *unc-103* (Garcia and Sternberg, 2003). Mutant *unc-103* males protrude their spicules in the absence of hermaphrodite stimulation. Some that do not protrude their spicules exhibit an abnormal prodding behavior when mating, interspersing long protractions without vulva penetration among short ones. These abnormal behaviors require the L-type calcium channel EGL-19, acetylcholine, and either the SPC neuron or the anal depressor muscle, which makes gap junctions onto the spicule protractors (Sulston *et al.*, 1980). Expression studies suggest that UNC-103 functions within neurons. ERG-like potassium channels are characterized by fast inactivation and recovery and presumably desensitize SPC and possibly other neurons (innervating the anal depressor), preventing them from misfiring during the prodding phase. This role is consistent with the function of this channel family in other animals: flies mutant in *sei/erg* channels have seizures; human mutants suffer heart arrhythmias.

XVI. Prospects

Among *C. elegans* behaviors, male-mating behavior offers several unique opportunities for learning more about higher level functions of nervous systems. Unlike other *C. elegans* behaviors, copulation by the male consists of a series of sub-behaviors, each sub-behavior contingent upon a particular pattern of sensory inputs. Dissecting how the nervous system carries out the monitoring and maintenance of a stable mode of behavior while sensory input remains within a range and switches to a new behavioral mode when change in sensory input exceeds a threshold, presents an interesting challenge. This dissectible pathway may help to reveal general principles of neural circuit design and function that are applicable throughout nervous systems. As a motivated behavior, sexual behavior is regulated and prioritized with other needs of the animal, such as feeding. With *C. elegans* mate-searching, a spontaneously-expressed behavior that seems different from a mere stimulus-response, this aspect is most apparent. However, copulation itself is dependent on conditions within the animal, as males deprived of food, though they may have both adequate energy and sperm, will not mate. Thus, male sexual behavior affords the opportunity to study how different overall modes of behavior that affect the entire nervous system are adaptively integrated.

Approaching these questions will require not only all of the sophistication provided by the genetic and molecular tools available in *C. elegans*, but also knowledge of the structure of the male nervous system. A renewed effort to provide complete connectivity data is justified by improvements in computer technology that have occurred since the first reconstructions in *C. elegans* were carried out on the hermaphrodite and male nervous systems in the 1970s (White, 1974). Results so far in a current project to reconstruct the male together with the earlier efforts reveal that many male-specific neurons are more highly branched than most hermaphrodite neurons (S. W. Emmons, M. Xu, D. Albertson, N. Thomson, D. Hall, 2005, unpublished results; Sulston *et al.*, 1980). While circuits and overall patterns of connectivity are undoubtedly constant from animal to animal, it appears unlikely that the precise pattern and location of individual synapses will be constant. Thus, mechanisms of wiring found in more complex animals, involving approximate connection followed by usage-dependent refinement, may occur in the wiring of the *C. elegans* male tail. Many neuronal interactions governing sexual behaviors and their integration with non-sexual, goal-oriented behaviors are likely to be humoral. Distinguishing humoral interactions from those involving synapses in circuits requires that the entire pattern of connectivity be known.

Finally, male sexual behavior, like most other *C. elegans* behavior, is innate; it is determined entirely by the genetic endowment. How this comes about in detail remains to be determined and forward genetics remains a powerful approach. Studies such as those uncovering roles of polycystins and ERG family channels show how genetic analysis of *C. elegans* male-mating behavior can lead to insights into conserved molecular mechanisms. Ultimately, we wish to trace behavior to its encoded form in the deoxyribonucleic acid sequence. With sexual behavior there is the advantage that the initial genetic signal, the gender-determination pathway (in *C. elegans*, the activity of the Zn-finger transcription factor TRA-1), is known. It seems reasonable to expect that unraveling the basis of sexual dimorphism in the *C. elegans* nervous system will provide insights applicable to all animals.

Acknowledgments

I thank R. Garcia for discussions and for his insights into *C. elegans* male copulatory behavior, and R. Lints for Figure 2. I am grateful to D. Fitch and D. Portman for their comments on the manuscript. Work in this laboratory is supported by grants from the U.S. National Institutes of Health. SWE is the Siegfried Ullmann Professor of Molecular Genetics.

References

Baird, S. E., and Emmons, S. W. (1990). Properties of a class of genes required for ray morphogenesis in *Caenorhabditis elegans*. *Genetics* **126,** 335–344.

Baird, S. E., Fitch, D. H. A., Kassem, I. A. A., and Emmons, S. W. (1991). Pattern formation in the nematode epidermis: Determination of the spatial arrangement of peripheral sense organs in the *C. elegans* male tail. *Development* **113,** 515–526.

Barker, D. M. (1994). Copulatory plugs and paternity assurance in the nematode *Caenorhabditis elegans*. *Anim. Behav.* **48,** 147–156.

Barr, M. M., DeModena, J. K., Braun, D., Nguyen, C. Q., Hall, D. H., and Sternberg, P. W. (2001). The *Caenorhabditis elegans* autosomal dominant polycystic kidney disease gene homologs *lov-1* and *pkd-2* act in the same pathway. *Curr. Biol.* **11,** 1341–1346.

Barr, M. M., and Garcia, L. R. (2005). Male mating behavior. *In* "WormBook, The *C. elegans* Research Community," www.wormbook.org.

Barr, M. M., and Sternberg, P. W. (1999). A polycystic kidney-disease gene homologue required for male mating behaviour in *C. elegans*. *Nature* **401,** 386–389.

Chasnov, J. R., and Chow, K. L. (2002). Why are there males in the hermaphrodite species *Caenorhabditis elegans*? *Genetics* **160,** 983–994.

Chow, K. L., and Emmons, S. W. (1994). HOM-C/Hox genes and four interacting loci determine the morphogenetic properties of single cells in the nematode male tail. *Development* **120,** 2579–2593.

Chow, K. L., Hall, D. H., and Emmons, S. W. (1995). The *mab21* gene of *Caenorhabditis elegans* encodes a novel protein required for choice of alternate cell fates. *Development* **121,** 3615–3626.

Coates, J. C., and de Bono, M. (2002). Antagonistic pathways in neurons exposed to body fluid regulate social feeding in *Caenorhabditis elegans*. *Nature* **419,** 925–929.

Corey, D. P. (2003). New TRP channels in hearing and mechanosensation. *Neuron* **39,** 585–588.

de Bono, M., Tobin, D. M., Davis M. W., Avery, L., and Bargmann, C. I. (2002). Social feeding in *Caenorhabditis elegans* is induced by neurons that detect aversive stimuli. *Nature* **419,** 899–903.

de Bono, M., and Bargmann, C. I. (1998). Natural variation in a neuropeptide Y receptor homolog modifies social behavior and food response in *C. elegans. Cell* **94,** 679–689.

Emmons, S. W., and Sternberg, P. W. (1997). Male development and mating behavior. *In* "*C. elegans* II" (D. L. Riddle, T. Blumenthal, B. J. Meyer, and J. R. Priess, Eds.), pp. 295–334. Cold Spring Harbor Laboratory Press, Cold Spring Harbor, New York.

Fitch, D. H. A. (2000). Evolution of "Rhabditidae" and the male tail. *J. Nematology* **32,** 235–244.

Garcia, L. R., Mehta, P., and Sternberg, P. W. (2001). Regulation of distinct muscle behaviors controls the *C. elegans* male's copulatory spicules during mating. *Cell* **107,** 777–788.

Garcia, L. R., and Sternberg, P. W. (2003). *Caenorhabditis elegans* UNC-103 ERG-like potassium channel regulates contractile behaviors of sex muscles in males before and during mating. *J. Neurosci.* **23,** 2696–2705.

Green, C. D. (1980). Nematode sex attractants. *Helminthological Abstracts, Series B* **49,** 81–93.

Hedgecock, E. M. (1976). The mating system of *Caenorhabditis elegans*: Evolutionary equilibrium between self- and cross-fertilization in a facultative hermaphrodite. *American Naturalist* **110,** 1007–1012.

Hodgkin, J. (1983). Male phenotypes and mating efficiency in *Caenorhabditis elegans. Genetics* **103,** 43–64.

Hodgkin, J. (1992). Genetic sex determination mechanisms and evolution. *BioEssays* **14,** 253–261.

Hodgkin, J., and Doniach, T. (1997). Natural variation and copulatory plug formation in *Caenorhabditis elegans. Genetics* **146,** 149–164.

Jiang, M., Ryu, J., Kiraly, M., Duke, K., Reinke, V., and Kim, S. K. (2001). Genome-wide analysis of developmental and sex-regulated gene expression profiles in *Caenorhabditis elegans. PNAS* **98,** 218–223.

Kleemann, G. (2005). "Mating Behavior and Intersexual Mating Interactions in the Nematode *Caenorhabditis elegans.*" University of Nebraska, Lincoln Nebraska.

Kupfermann, I., Kandel, E. R., and Iversen, S. (2000). Motivational and addictive states. *In* "Principles of Neural Science" (E. R. Kandel, J. H. Schwarz, and T. M. Jessell, Eds.), pp. 998–1013. Elsevier, New York.

Lee, D. L. (2002). Behavior. *In* "The Biology of Nematodes" (D. L. Lee, Ed.), pp. 369–387. Taylor and Francis, London.

Lints, R., and Emmons, S. W. (1999). Patterning of dopaminergic neurotransmitter identity among *Caenorhabditis elegans* ray sensory neurons by a TGFb family signaling pathway and a *Hox* gene. *Development* **126,** 5819–5831.

Lints, R., and Emmons, S. W. (2002). Regulation of sex-specific differentiation and mating behavior in *C. elegans* by a new member of the DM domain transcription factor family. *Genes Dev.* **16,** 2390–2402.

Lints, R., Jia, L., Kim, K., Li, C., and Emmons, S. W. (2004). Axial patterning of *C. elegans* male sensilla identities by selector genes. *Dev. Biol.* **269,** 137–151.

Lipton, J., Kleeman, G., Ghosh, R., Lints, R., and Emmons, S. W. (2004). Mate searching in *Caenorhabditis elegans*: A genetic model for sex drive in a simple invertebrate. *J. Neurosci.* **24,** 7427–7434.

Liu, K., and Sternberg, P. W. (1995). Sensory regulation of male mating behavior in *Caenorhabditis elegans. Neuron* **14,** 79–89.

Loer, C. M., and Kenyon, C. J. (1993). Serotonin-deficient mutants and male mating behavior in the nematode *Caenorhabditis elegans. J. Neurosci.* **13,** 5407–5417.

Matsuda, M., Nagahama, Y., Shinomiya, A., Sato, T., Matsuda, C., Kobayashi, T., Morrey, C. E., Shibata, N., Asakawa, S., Shimizu, N., Hori, H., Hamaguchi, S., and Sakaigami, M. (2002). *DMY* is a Y-specific DM-domain gene required for male development in the medaka fish. *Nature* **417,** 559–563.

Meyer, B. J. (1997). Sex determination and X chromosome dosage compensation. *In* "*C. elegans* II" (D. L. Riddle, T. Blumenthal, B. J. Meyer, and J. R. Priess, Eds.), pp. 209–240. Cold Spring Harbor Laboratory Press, Cold Spring Harbor, New York.

Nauli, S. M., Alenghat, F. J., Luo, Y., Williams, E., Vassilev, P., Li, X., Elia, A. E., Lu, W., Brown, E. M., Quinn, S. J., Ingber, D. E., and Zhou, J. (2003). Polycystins 1 and 2 mediate mechanosensation in the primary cilium of kidney cells. *Nat. Genet.* **33,** 129–137.

Nguyen, C. Q., Hall, D. H., Yang, Y. L., and Fitch, D. H. A. (1999). Morphogenesis of the *Caenorhabditis elegans* male tail tip. *Dev. Biol.* **207,** 86–106.

Perry, R. N., and Aumann, J. (1998). Behaviour and sensory responses. *In* "The Physiology and Biochemistry of Free-Living and Plant-Parasitic Nematodes" (R. N. Perry and D. J. Wright, Eds.), pp. 75–102. CABI, Wallingford.

Portman, D. S., and Emmons, S. W. (2004). Identification of *C. elegans* sensory ray genes using whole-genome expression profiling. *Dev. Biol.* **270,** 499–512.

Raymond, C. S., Murphy, M. W., O'Sullivan, M. G., Bardwell, V. G., and Zarkower, D. (2000). *Dmrt1*, a gene related to worm and fly sexual regulators, is required for mammalian testis differentiation. *Genes Dev.* **14,** 2587–2595.

Raymond, C. S., Shamu, C. E., Shen, M. M., Seifert, K. J., Hirsch, B., Hodgkin, J., and Zarkower, D. (1998). Evidence for evolutionary conservation of sex-determining genes. *Nature* **391,** 691–695.

Salser, S., and Kenyon, C. J. (1996). A *C. elegans* Hox gene switches on, off, on and off again to regulate proliferation, differentiation and morphogenesis. *Development* **122,** 1651–1661.

Shen, M. M., and Hodgkin, J. (1988). *mab-3,* a gene required for sex-specific yolk protein expression and a male-specific lineage in *C. elegans. Cell* **54,** 1019–1031.

Simon, J. M., and Sternberg, P. W. (2002). Evidence of a mate-finding cue in the hermaphrodite nematode *Caenorhabditis elegans. Proc. Natl. Acad. Sci. USA* **99,** 1598–1603.

Stewart, A. D., and Phillips, P. C. (2002). Selection and maintenance of androdioecy in *Caenorhabditis elegans. Genetics* **160,** 975–982.

Sudhaus, W., and Fitch, D. (2001). Comparative studies on the phylogeny and systematics of Rhabditidae (Nematoda). *J. Nematology* **33,** 1–72.

Sulston, J. E., Albertson, D. G., and Thomson, J. N. (1980). The *Caenorhabditis elegans* male: Postembryonic development of nongonadal structures. *Dev. Biol.* **78,** 542–576.

Sulston, J. E., and Horvitz, H. R. (1977). Post-embryonic cell lineages of the nematode, *Caenorhabditis elegans. Dev. Biol.* **56,** 110–156.

Troemel, E. R., Chou, J. H., Dwyer, N. D., Colbert, H. A., and Bargmann, C. I. (1995). Divergent seven transmembrane receptors are candidate chemosensory receptors in *C. elegans. Cell* **83,** 207–218.

Ward, S., and Carrel, J. S. (1979). Fertilization and sperm competition in the nematode *Caenorhabditis elegans. Dev. Biol.* **73,** 305–321.

White, J. G. (1974). Computer aided reconstruction of the nervous system of *C. elegans.* Ph.D. Thesis. University of Cambridge, England.

White, J. G., Southgate, E., Thomson, J. N., and Brenner, S. (1986). The structure of the nervous system of *Caenorhabditis elegans. Philos. Trans. R Soc. Lond. B. Biol. Sci.* **314,** 1–340.

Yi, W., Ross, J. M., and Zarkower, D. (2000). *mab-3* is a direct *tra-1* target gene regulating diverse aspects of *C. elegans* male sexual development and behavior. *Development* **127,** 4469–4480.

Zhang, Y., and Emmons, S. W. (1995). Specification of sense-organ identity by a *Caenorhabditis elegans Pax-6* homologue. *Nature* **377,** 55–59.

THE MOTOR CIRCUIT

Stephen E. Von Stetina,* Millet Treinin,† and David M. Miller, III*

*Department of Cell and Developmental Biology, Vanderbilt University
Nashville, Tennessee

†Department of Physiology, Hebrew University-Hadassah Medical School
Jerusalem, Israel

Caenorhabditis elegans motor neurons control a range of activities including locomotion, foraging, defecation, and gender-specific functions. In this chapter, we focus primarily on motor neurons that regulate body movement, with particular emphasis on those in the ventral nerve cord (VNC). We describe the basic

INTERNATIONAL REVIEW OF
NEUROBIOLOGY, VOL. 69
DOI: 10.1016/S0074-7742(05)69005-8

architecture and development of the motor circuit, genes that specify motor neuron fates, and models of how the motor circuit controls locomotion. We identify surprising similarities between the structure and development of the nematode and vertebrate axial nerve cords and speculate about the potential roles of conserved families of transcription factors in the evolution of these motor circuits.

I. Introduction

Animal movement depends on the coordinated activities of the nervous system and the muscles it controls. Muscles that regulate locomotion and internal organ function are innervated by motor neurons residing in an axial nerve cord. Command signals enter this network from anteriorly located ganglia (e.g., brain) that extend processes into the nerve cord to synapse with local motor circuit neurons. Additional motor neurons in the "head" region may comprise a separate network for muscle functions unique to the anterior end of the animal (e.g., control of mouth parts). Although these arrays are larger and more complex in vertebrates than in non-vertebrates, this basic architecture and arrangement of motor circuit components is generally preserved in bilaterally symmetrical animals. Thus, it should be possible to exploit the simplicity and experimental manipulability of model organisms such as *C. elegans* to define fundamentally important features of motor circuit development and function.

II. An Overview of *C. elegans* Motor Circuit Anatomy

Motor neurons, defined as neurons that synapse with muscle cells, comprise over one-third of all neurons in the *C. elegans* nervous system (113 of 302) (White *et al.,* 1986). A majority of these motor neurons (80) are distributed along the ventral nerve cord (VNC) and in ganglia at either end (White *et al.,* 1976) (Fig. 1). A subset of the VNC motor neurons extends circumferential processes (commissures) dorsally to establish a separate dorsal nerve cord. Together, these motor neurons regulate dorsal/ventral body bends that drive locomotion. Other motor neurons, located in the head, facilitate a wider range of movements at the anterior end. Motor neuron activity in the axial nerve cords depends on input from "command interneurons" which extend axonal processes into the cord from head or tail ganglia. Motor neurons, in turn, provide output to longitudinally arrayed rows of muscle cells attached to the cuticular exoskeleton that envelops the animal (Fig. 2).

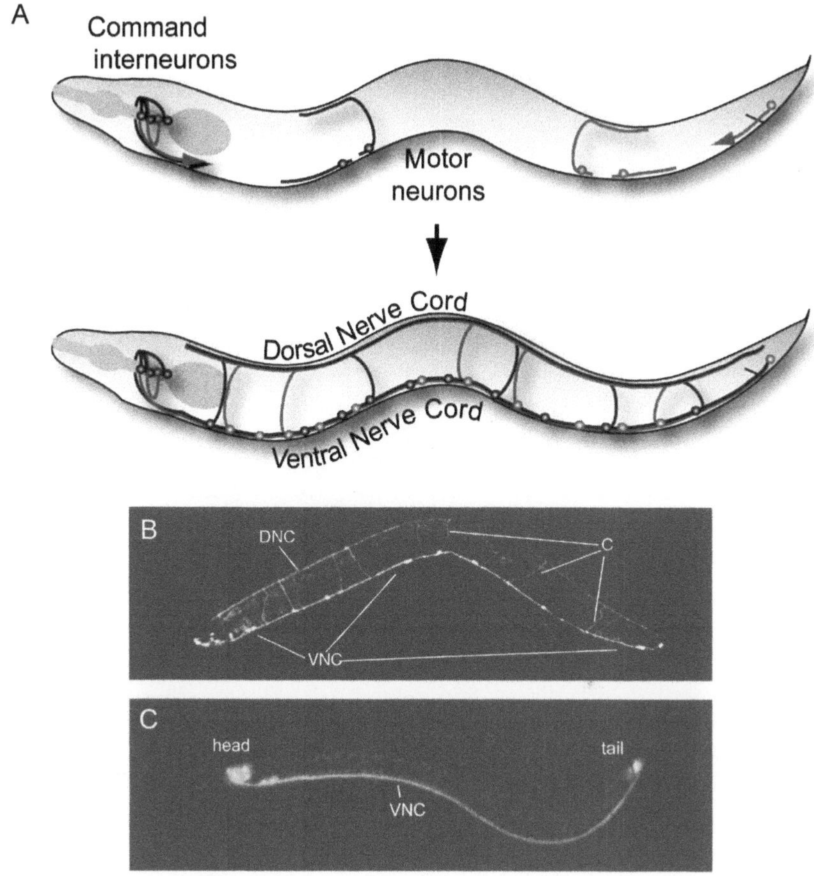

FIG. 1. The *C. elegans* motor circuit. A. Command interneurons in the head and tail extend processes into the ventral nerve cord to synapse with specific classes of motor neurons. "Red" and "Blue" circuits denote separate sets of command interneurons and motor neuron partners. Selected motor neurons shown here are presynaptic to adjacent body muscle cells in either the ventral or dorsal nerve cords. B. GFP driven by the *unc-47* promoter labels the GABA circuit with motor neurons extending processes in the ventral nerve cord (VNC) and to the dorsal nerve cord (DNC) via commissures (C) (reprinted from Francis, 2003). C. A subset of neurons, including the command interneurons AVA, AVD, AVE (head) and PVC (tail) are marked with the *nmr-1*::GFP reporter to show projections into the VNC (adapted from Francis, 2003). (See Color Insert.)

A. Muscle

The adult body musculature is composed of 95 cells (Moerman and Fire, 1997). Although nematode muscle cells do not undergo membrane fusion as is the case in insect and mammals, they are tightly apposed and electrically coupled

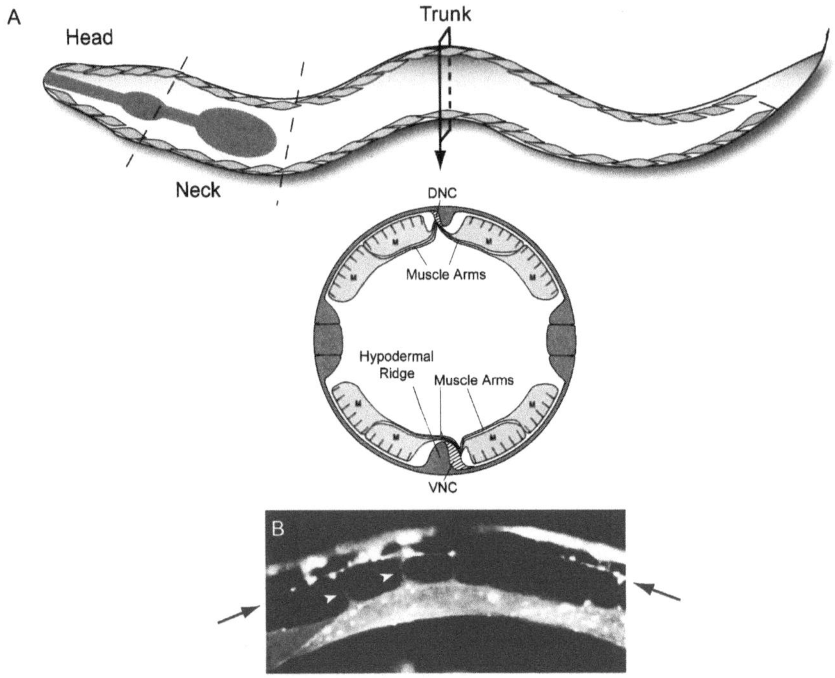

FIG. 2. *C. elegans* body-wall muscle. A. Spindle-shaped body wall muscle cells can be grouped in three regions (head, neck, trunk) based on innervation by motor neurons in the VNC (see text for details). A cross-section through the trunk shows muscle arms projecting from body wall muscle cells (M) to the ventral and dorsal cords. (Adapted from White *et al.*, 1986). B. Ventral view of adult hermaphrodite. Membrane-bound YFP illuminates selected muscle cells in ventral quadrants and muscle arms (arrowheads) projecting to the VNC (arrows). (Adapted from Dixon and Roy, 2005).

by prominent gap junctions (White *et al.*, 1986). Body muscle cells are situated in four quadrants flanking the nerve cords. Each quadrant is comprised of two rows of muscle cells. These are further subdivided into three groups—head, neck and body—on the basis of motor neuron innervation along the anterior/posterior axis (Fig. 2A). The 16 head muscle cells are the most anterior and receive input from neurons in the head and nerve ring. The 16 neck muscles accept mixed innervation from both head and ventral cord motor neurons. The remaining 63 body muscles synapse exclusively with VNC motor neurons. Thus, the head versus body motor circuits constitute two separate neuromuscular units, with each group of muscles receiving inputs from distinct sets of motor neurons. These differences are correlated at least in part with the developmental origins of the motor neurons that innervate these muscles; head motor neurons are almost exclusively derived in the embryo whereas ventral cord motor neurons are generated in both embryonic and larval stages (Table I). Key regulators

TABLE I
MOTOR NEURON CLASSES INVOLVED IN MOVEMENT

Class	# Neurons	Origin	Muscles innervated	Neurotransmitter[b]
Head neurons				
IL1	6	E	Head	?
RIM	2	E	Neck	ACh
RIV	2	E	Ventral neck	?
RMD	6	E	Head, neck	ACh
RME	4	E	Head, neck	GABA
RMF	2	L	Head	?
RMG	2	E	Neck	?
RMH	2	L	Head	?
SMB	4	E	Head, neck	ACh
SMD	4	E	Head, neck	ACh
URA	4	E	Head	ACh
Ventral cord neurons				
DA	9	E	Neck[a], dorsal body	ACh
DB	7	E	Neck[a], dorsal body	ACh
DD	6	E	Neck[a], dorsal body	GABA
VA	12	L	Neck[a], ventral body	ACh
VB	11	L	Neck[a], ventral body	ACh
VC	6	L	Neck[a], ventral body	ACh
VD	13	L	Neck[a], ventral body	GABA
AS	11	L	Neck[a], ventral body	ACh

E = embryonic; L = larval. Adapted from White *et al.* (1986).
[a]Only anterior members of each motor neuron class (e.g., DA1) innervate neck muscle cells.
[b]Neurotransmitter assignments from www.wormbase.org.

of motor neuron fate also differ between these two broad motor neuron classes which suggest that motor neuron characteristics are largely defined in these cases by separate genetic programs (see the following text). Differential gene expression in body muscle cells has been reported but the potential contributions of these differences to muscle development and function have not been defined (Christensen *et al.,* 2002; Colavita *et al.,* 1998; Moerman and Fire, 1997).

B. MUSCLE ARMS PROJECT TO MOTOR NEURON TARGETS

A neuromuscular junction (NMJ) is defined as a specialized chemical synapse in which at least one post-synaptic partner is muscle. In mammals, the NMJ is distinguished from central synapses by the presence of a basal lamina that bisects the synaptic cleft between the motor axon and muscle target (Sanes and

Lichtman, 1999). The basal lamina is also evident at the nematode NMJ as *C. elegans* neurons are sandwiched between the skin ("hypodermis") and a layer of extracellular matrix material (Moerman and Fire, 1997). In insects and vertebrates, NMJs arise at the tips of motor axons that project toward muscle targets. In contrast, in nematodes, the muscle cells extend elongated processes called "arms" to synapse with motor neurons (Fig. 2B). Although this unusual architecture would appear to reflect a fundamental divergence of NMJ developmental programs between mammals and nematodes, recent work has shown that muscle cells in flies and in some vertebrates extend transient "myopodia" toward motor neurons during NMJ formation (Misgeld *et al.*, 2002; Ritzenthaler and Chiba, 2003).

In the embryo, muscle arms are generated by a passive mechanism in which the myoblast soma initially contacts the presynaptic motor neuron and then migrates away leaving a cytoplasmic bridge to the NMJ in its wake (C. R. Norris, I. A. Bazykina, E. M. Hedgecock, D. H. Hall, 1997, personal communication). In contrast, larval muscle arms are established by an active process in which membrane protrusions extend from body muscle cells toward motor neurons targets in the VNC (Dixon and Roy, 2005). The directed outgrowth of larval muscle arms is suggestive of a steering mechanism that depends on chemoattractive cues from the motor neurons. This model is consistent with other observations. For example, in certain axon guidance mutants, some motor neuron commissures stop short of the dorsal nerve cord and instead enter sublateral processes. In these cases, ectopic NMJs are created by muscle arms that extend downward from dorsal muscle cells toward these displaced motor axons (Hedgecock *et al.*, 1990). Although the identity of the chemoattractive signal is unknown, it could originate from synaptic vesicles; motor neuron soma are aberrantly targeted by muscle arms when synaptic vesicle transport into the axon is blocked in *unc-104* kinesin mutants (Hedgecock and Hall, 1991). Additional evidence suggests that muscles may also provide attractive cues to motor neurons. When selected body muscle cells are killed with a laser, adjacent D-class motor neurons (DD/VD, see the following text) often branch across these gaps toward distal muscle cells. In addition, consistent with the hypothesis that neurons also provide an attractive signal, adjacent muscles project into the ablated domain toward D-type processes (Plunkett *et al.*, 1996).

C. The Head Motor circuit

NMJs with head and neck body muscles (Fig. 2) are located on the inside surface of the nerve ring, situated between motor neuron processes and the glial-like GLR cells (which may be guideposts for muscle arms) (White *et al.*, 1976). Muscle arms that project to these locations are precisely placed as are the presynaptic axons that encircle them. In this arrangement, each muscle cell quadrant is discretely connected to neuronal processes in the adjacent quarter of the nerve ring. In addition, inputs to each of the two rows of muscle cells within each quadrant are also separate. Thus, NMJs within the nerve ring map to 8 distinct regions, one for each row of body muscle cells (White *et al.*, 1976). This organization may account for the wide range of movement for the head as opposed to the body which is limited to dorsal/ventral flexures. Nerve ring inputs to muscle cells are provided by 14 separate classes of neurons (Table I) (White *et al.*, 1986). Individual neurons in this complex circuit are believed to modulate head muscle activity in one or more of a wide range of behaviors: forward locomotion, foraging (sweeping of the head from side-to-side within a narrow arc while feeding), and avoidance responses (reflexive movement away from a touch stimulus) (Driscoll and Kaplan, 1997). For example, laser ablations and mutant analyses have established that the RME motor neurons are required for normal head movements during foraging (McIntire *et al.*, 1993b). Other nerve ring neurons with inputs to muscles (e.g., IL-1, RMD) may function as combined sensory and motor neurons to modulate head movement in response to external stimuli (D. Elkes and J. Kaplan, 1993, personal communication). A few of the NMJs in the nerve ring are dyadic, that is one motor neuron synapses onto two post-synaptic targets (muscle and another neuron). These post-synaptic motor neurons (RMD, SMD, and RME) 'intercept' an NMJ by projecting a thin dendritic shaft into the synapse. These intercepted inputs are contra-lateral to their outputs, suggesting that these neurons may be cross-inhibitors (White *et al.*, 1986). This model is consistent with the observation that at least one of these motor neuron classes, the RMEs, express the inhibitory neurotransmitter GABA (McIntire *et al.*, 1993b).

D. The Ventral Nerve Cord (VNC)

The VNC is comprised of a dense bundle of neuronal processes and motor neuron soma wedged between the ventral hypodermal ridge and an overlapping sheet of basal lamina (Fig. 3A) (White *et al.*, 1976, 1986). The VNC is asymmetric with most processes displaced to the right side of the hypodermal ridge. These include all ventral cord motor neurons (see the following text) as well as their presynaptic partners, the command interneurons (AVA, AVD, AVE, AVB, and PVC) which extend axonal processes into the VNC from ganglia at either end. Neuronal processes are oriented along the anterior/posterior axis such that synapses are made *en passant* between adjacent parallel processes. Motor neuron processes are largely unbranched and course between different neighborhoods within the cord to accommodate input from the command interneurons in the middle of the fascicle and output to muscle arms touching the periphery. The

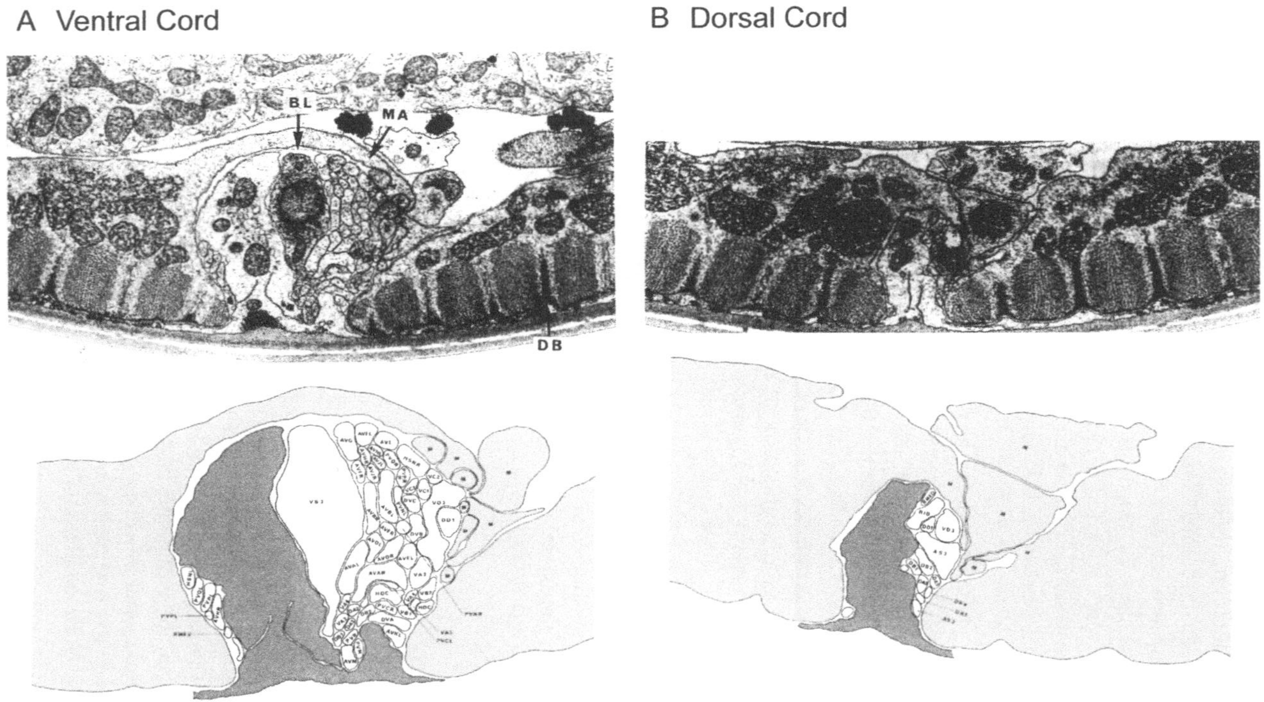

FIG. 3. Ventral and dorsal nerve cords. A. EM cross-section through the ventral nerve cord (VNC) of an adult hermaphrodite. Note that the right side of the cord contains most of the neuronal processes, including those from all motor neurons. Muscle cells (M) from both left and right quadrants project muscle arms (MA) to the right side of the cord to make neuromuscular junctions across the basal lamina (BL). Dense bodies (DB) anchor muscle cell sarcomeres to the cuticular exoskeleton. B. EM cross-section through the dorsal nerve cord (DNC) of an adult hermaphrodite, showing fewer processes than the VNC. Lower panels show labeled outlines of VNC and DNC neuronal processes. The hypodermal ridge (dark gray) is continuous with a thin sheet of ectoderm (not shown) that envelops the animal (see Fig. 2). (Adapted from White *et al.*, 1986.)

dorsal nerve cord (DNC) contains fewer processes (Fig. 3B). Most of these are commissural axons from ventral cord motor neurons joined by a limited set of neurons in the head (e.g., RMED) and tail (e.g., PDA) that extend short processes into the bundle.

E. VNC Motor Neurons

Body muscles in the trunk region (see the previous text) are innervated by motor neurons located in the VNC and in ganglia at each end. These motor neurons can be grouped into eight distinct classes: DA, VA, DB, VB, DD, VD, VC, and AS (Fig. 4). Four of these innervate ventral muscles (VA, VB, VD, VC) whereas the others (DA, DB, DD, AS) extend commissures to synapse with dorsal muscle cells (White *et al.*, 1976, 1986). The active zones for the members of each motor neuron class do not overlap although active zones from different classes of motor neurons are presynaptic to common sets of muscle arms. Synaptic inputs from command interneurons are restricted to the VNC and obey highly reproducible and specific patterns of innervation. These connections include both gap junctions and chemical synapses. Motor neurons of a given class are also electrically coupled via gap junctions in both the DNC and VNC. Subgroupings of the motor neuron classes can be recognized on the basis of shared morphological features and common presynaptic inputs. For example, A-type motor neurons (DA, VA) extend anteriorly directed axonal processes and synapse with the command interneurons AVA (gap junction and chemical synapse), AVD, and AVE (chemical synapse) whereas axons from B-type motor neurons (DB, VB) adopt a posterior trajectory and receive input from PVC (chemical synapse) and AVB (gap junction). The D-type motor neurons (DD, VD) are strictly postsynaptic to other motor neurons and use the neurotransmitter GABA.

F. Development of the Ventral Cord Motor Circuit

Motor neurons are generated during two distinct developmental periods, initially in the embryo and also later after hatching in the first larval stage. The lineal origin and time of birth for each of these cells are highly reproducible and well-defined (Sulston, 1976; Sulston and Horvitz, 1977; Sulston *et al.*, 1983).

1. *The Embryonic Motor Circuit*

DA, DB, and DD motor neurons arise from distinct lineages to occupy specific ventral locations about midway through embryonic development (Sulston *et al.*, 1983). All three classes extend commissures to the dorsal cord.

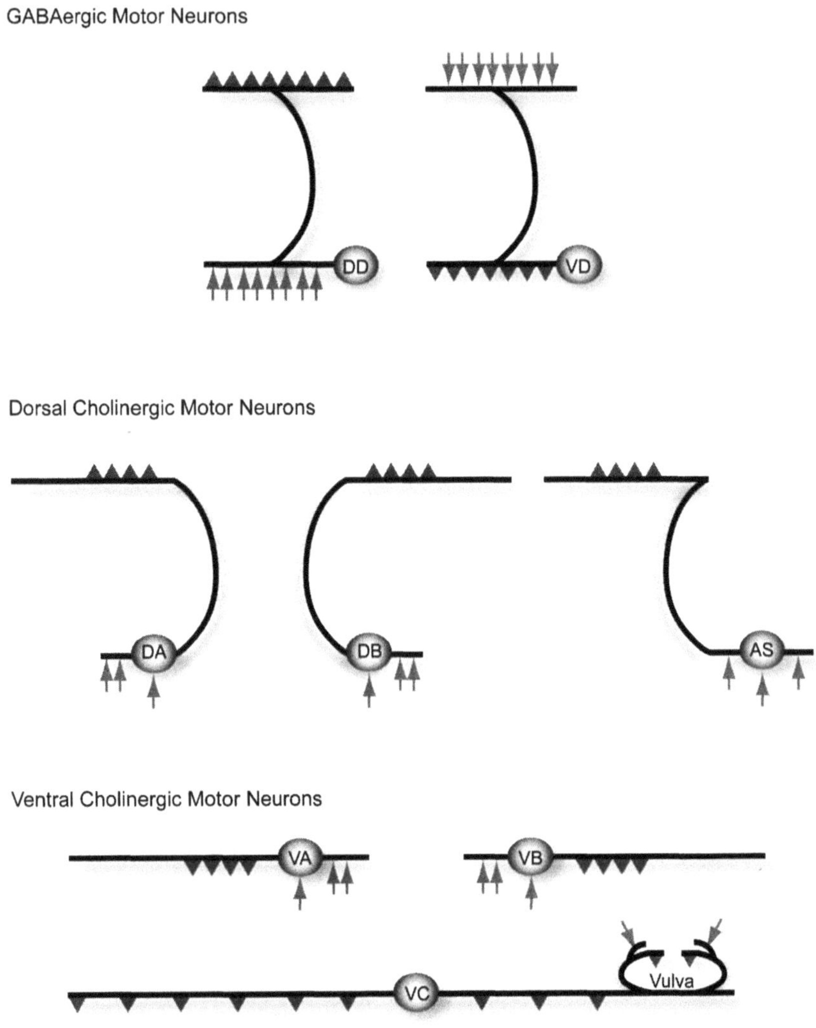

FIG. 4. Morphology of adult ventral cord motor neurons. Representative members of each motor neuron class are shown here. Note that five of the motor neuron classes (DA, DB, DD, AS, VD) extend commissures to the DNC whereas three (VA, VB, VC) are restricted to the VNC. Each motor neuron is representative of its class although minor differences in morphology do occur within each class. For example, in contrast to other VC class motor neurons, VC4 and VC5 do not make NMJs with body wall muscle as indicated in this diagram. Green arrows depict input from other neurons. Red triangles denote NMJs with body wall muscle cells. (See Color Insert.)

The DA axons turn anteriorly upon reaching the dorsal side, the DBs turn posteriorly, and the DDs send out projections in both directions (Fig. 4). At hatching, DA and DB neurons innervate dorsal muscle whereas the DDs synapse with ventral muscle (White *et al.,* 1978). (DD motor neurons reverse polarity to innervate dorsal muscles after hatching; see the following text.) DA and DB motor neurons are excitatory and express acetylcholine (Rand and Nonet, 1997). Dorsal processes from the inhibitory and GABAergic DD motor neurons 'intercept' DA and DB cholinergic NMJs in the dorsal cord to modulate locomotion (McIntire *et al.,* 1993b).

All but one of the embryonic ventral cord motor neurons arise from the ABp lineage (Fig. 5) (DB2 is derived from ABa) (Sulston *et al.,* 1983). The lineal origins of these motor neurons reflect the overall bilateral symmetry of the nervous system with approximately equal numbers of motor neurons arising from right and left sides. In addition, identical lineages on the right and left sides typically give rise to motor neurons of the same class (with some exceptions, e.g., DA8 versus DB5). The six DD motor neurons are essentially clonal in origin with three DD neurons arising from a common precursor cell on both sides. Subsets of the DA motor neurons (e.g., DA1, DA3, an DA5) are also clonally derived but all seven of the DB motor neurons are generated from independent lineages. Although it is clear that embryonic motor neuron subtypes are derived from disparate lineages, all of these patterns of cell division result in the proper placement of neuroblasts along the ventral side of the developing embryo. (This outcome is achieved in a much more orderly fashion for larval motor neuron classes which arise from a linear array of ventrally placed embryonic blast cells [see the following text].) Mechanisms governing embryonic motor neuron fates are largely undefined but recent evidence suggests that these cells may adopt specific fates in the absence of cues from their neighbors. Primary cultures established from mitotic embryonic blastomeres give rise to the expected fraction of motor neurons from each class. In addition, these motor neurons express predicted class-specific marker genes *in vitro* (Christensen *et al.,* 2002).

In contrast to this apparent cell autonomous mode of setting motor neuron class identity, the placement of motor neuron soma, process outgrowth, and synaptic connectivity are context dependant. All of the embryonic motor neuron cell bodies come to rest on the right side of the ventral hypodermal ridge. A single pioneering axon from the anteriorly located AVG neuron extends through this area to mark the location of the VNC. Notably, the AVG axon reaches the posterior end of the cord before motor neuron outgrowth begins (Durbin, 1987). AVG is likely to provide cues that govern motor neuron outgrowth as its removal by laser ablation at an earlier stage results in the misplacement of DD ventral processes and the extension of DA and DB commissures from the wrong side (Durbin, 1987; Hutter, 2003). Commissures grow out simultaneously from all

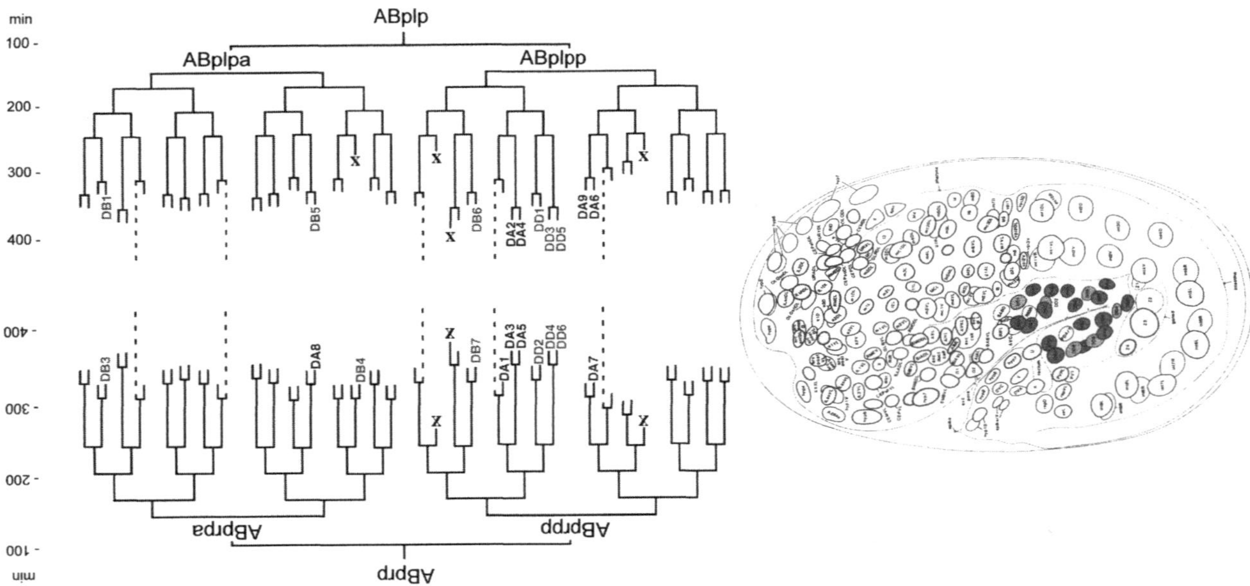

FIG. 5. Embryonic ventral cord motor neurons are derived from specific cell lineages. A. All but one of the embryonic VNC motor neurons are derived from two embryonic precursor cells, ABplp and ABprp (DB2 arises from ABa). The ABprp lineage is inverted to feature the overall bilateral symmetry of these cell divisions. The Y-axis refers to minutes after fertilization (at 20 C). (Adapted from Hallam et al., 2000.) B. Diagram of medial/left hand side of comma stage embryo to demonstrate the placement of the embryonic motor neurons along the ventral surface. (adapted from Sulston *et al.*, 1983.) (See Color Insert.)

three embryonic motor neuron classes; their dorsal trajectories are defined by a combination of repulsive signals (UNC-6, slit) presumptively provided by ventral ectoderm (Chang *et al.,* 2004; Wadsworth and Hedgecock, 1992) and attractive cues (e.g., UNC-129) expressed by dorsal muscle cells (Colavita *et al.,* 1998). Command interneurons enter the ventral cord later, after motor neuron outgrowth is complete, to synapse with specific targets. The specific mechanisms that define target recognition in this circuit are unknown.

2. *Generation of Larval Motor Neurons*

Five classes of motor neurons (VA, VB, VC, AS, and VD) are added to the VNC after hatching (Sulston, 1976; Sulston and Horvitz, 1977). All of these motor neurons arise from 13 blast cells or "P-cells" set aside during embryonic development. Toward the end of the L1 stage, six pairs of P-cells move downward from the left and right lateral hypodermis to the ventral midline (Fig. 6). This process is highly ordered as cytoplasmic arms emanating from these cells guide migrating nuclei to the ventral surface (Starr *et al.,* 2001). The 13th P-cell ("W") is already in place at the anterior end of the VNC at hatching. The migrating P-cells are numbered P1 to P12 according to their final locations in the queue. Upon reaching their destinations, the P-cells begin dividing along the A/P axis. Cell division is initiated at the anterior end such that postembryonic motor neurons are generated in an anterior-posterior wave of differentiation. The anterior daughter of the first division, Pn.a, undergoes a series of stereotypical divisions to produce all 53 of the postembryonic VNC motor neurons whereas posterior Pn.p daughters join the hypodermis or contribute to the vulva in the midbody region. The central six Pn.a cells, P3.a to P8a, generate identical lineages with each one producing single representatives of the five larval motor neuron classes. For example, the most anterior daughter, Pn.aaaa, is always a VA motor neuron whereas its posterior sister, Pn.aaap, gives rise to a VB. Deviations from this pattern, however, emerge at each end of the cord (Fig. 6). For example, the most anterior daughters of W and P1 are AVF interneurons rather than VAs. In another case, Pn.aap cells arising from the distal P1.a to P3.a and P9.a to P12.a precursors undergo programmed cell death. The net result is that only six VC motor neurons are generated in the midbody region to innervate the hermaphrodite vulva. In males, which lack this egg-laying organ, these cell deaths do not occur and the Pn.app cells adopt an alternative fate to give rise to male-specific motor neurons. Other differences in posterior lineages generate additional neurons in the male tail. Laser ablation experiments suggest that local signals could bias P-cell fates in these distal regions. For example, ablation of P1 results in P2 adopting the P1 lineage (i.e., generates AVF versus VA) (J. White, 2005, personal communication). Similar transformations in neuronal fates are observed with the removal of posteriorly located P-cells in the male (Sulston and White, 1980). These experiments evoke the concept of "equivalence groups" in which

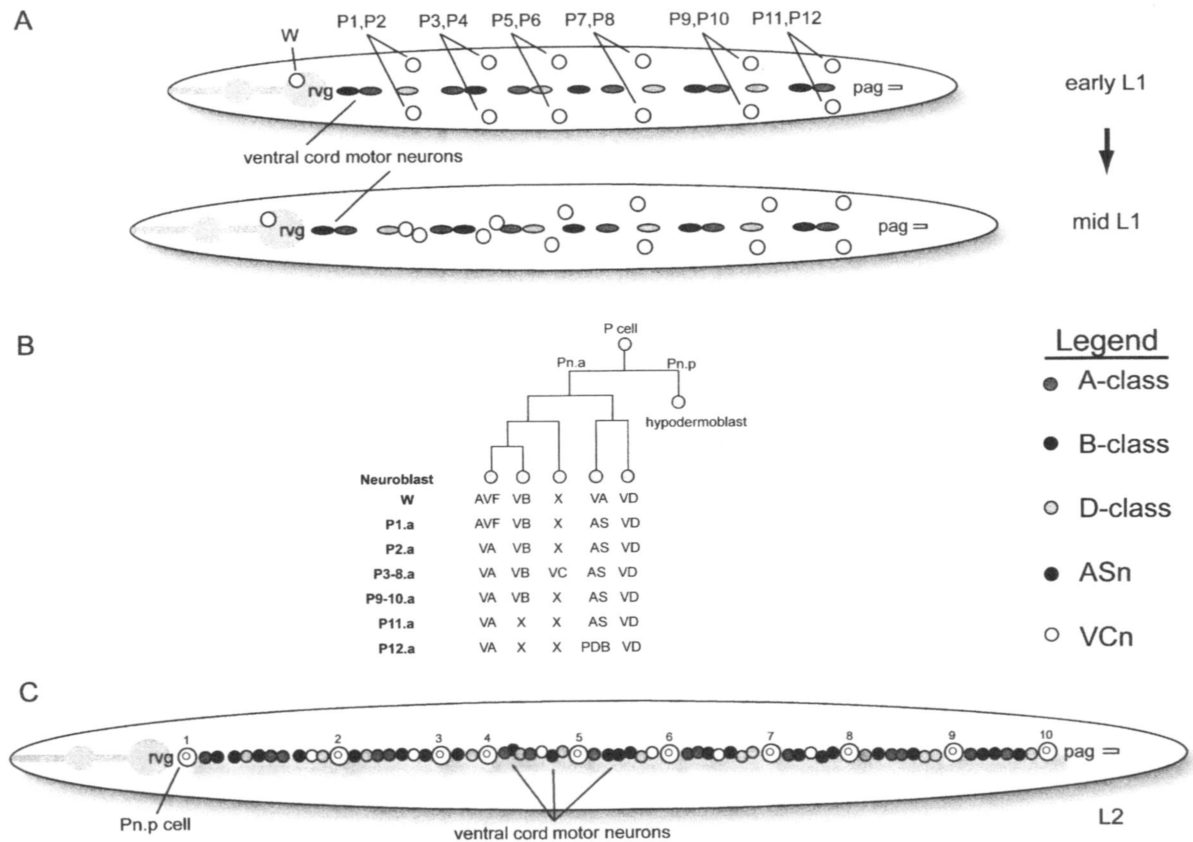
A
W
P1,P2
P3,P4
P5,P6
P7,P8
P9,P10
P11,P12
rvg
pag
early L1
ventral cord motor neurons
mid L1
B
P cell
Pn.a
Pn.p
hypodermoblast
Neuroblast
W AVF VB X VA VD
P1.a AVF VB X AS VD
P2.a VA VB X AS VD
P3-8.a VA VB VC AS VD
P9-10.a VA VB X AS VD
P11.a VA X X AS VD
P12.a VA X X PDB VD
Legend
A-class
B-class
D-class
ASn
VCn
C
1
2
3
4
5
6
7
8
9
10
Pn.p cell
ventral cord motor neurons
L2

adjacent blast cells are born with equal developmental potential but adopt different fates in response to biased inductive signals (Sulston *et al.,* 1983). Perhaps local cues from head and tail regions specify proximal motor neuron progenitors in a manner comparable to the induction of distinct motor neurons fates along the rostral-caudal axis of the vertebrate spinal cord. In this more complex motor circuit, graded signals emanating from an anteriorly located source and from adjacent mesodermal cells are key regulators of motor neuron differentiation (Appel and Eisen, 2003; Liu *et al.,* 2001). In the chick, motor neuron fates unique to these regions depend on differential expression of HOX family transcription factors (Dasen *et al.,* 2003). In this regard, it is interesting to note that the nematode HOX gene, *mab-5,* is selectively required for the adoption of a different fate (i.e., cell death) for the P11.aaap and P12.aaap progenitors versus their more anterior lineal homologs which differentiate as VB motor neurons (Salser *et al.,* 1993). At this point, however, the identities of potential inductive signals and how they might regulate differential gene expression in the *C. elegans* motor circuit are unknown.

3. *Reorganization of the Ventral Cord Motor Circuit During Larval Development*

The postembryonic motor neurons are integrated into the VNC soon after their birth in the L1 stage (Sulston, 1976; White *et al.,* 1976). Larval motor neuron processes must insinuate themselves into the pre-existing fascicle to contact appropriate presynaptic command interneurons and to make NMJs with new muscle arms that project to the dorsal and ventral cords. Existing contacts between adjacent processes are reorganized to accommodate the additional motor neurons. The VNC may be especially sensitive to perturbation during this period as specific diffusible signals, secreted Ig domain proteins, are produced in the VNC to prevent processes from slipping over the ventral hypodermal ridge to the opposite nerve tract (Aurelio *et al.,* 2002). Perhaps the most striking change in the motor circuit involves the embryonically derived DD motor neurons. In the newly hatched L1, DD motor neurons innervate ventral muscles. DA and DB motor neurons innervate dorsal muscles at dyadic synapse where they also

---

FIG. 6. Specific blast cells (P-cells) give rise to larval motor neurons in the ventral nerve cord. All animals are viewed from the ventral side. A. L1 larvae. P-cells migrate ventrally from lateral locations in an anterior-to-posterior wave in the early L1. (Adapted from Sulston 1976). B. The anterior P cell daughter (Pn.a) divides in the late L1 to generate larval motor neurons. Note deviations from stereotypical lineage of P3.a-P8.a for Pn.a cells located at anterior and posterior ends of the VNC (e.g. P1.a). "x" denotes a cell death (adapted from Chalfie and White, 1988). C. L2 larva showing the full complement of ventral cord motor neurons (small circles). Pn.p ectodermal blast cells (numbers) display prominent nucleoli and are reliable landmarks in the queue of ventral cord motor neuron nuclei. (Adapted from Sulston, 1976 and from Sulston and Horvitz, 1977). Additional motor neurons reside in ganglia at either end of the VNC are rvg (retrovesicular ganglion) and pag (preanal ganglion). (See Color Insert.)

provide output to DD dendrites in the dorsal nerve cord. Toward the end of the first larval stage, the DDs switch polarity so as to provide output to dorsal muscles while receiving input from newly generated ventral motor neurons (VA, VB). Remarkably, DD processes remain in place as presynaptic and postsynaptic specializations are reconstructed at opposite ends of the cell (White *et al.,* 1978). This process is developmentally regulated (Hallam and Jin, 1998) and correlated with the addition of VD motor neurons which now innervate the ventral muscle abandoned by the metamorphic DD motor neurons (see the following text). Interestingly, this reversal of DD motor neuron muscle output (i.e., ventral to dorsal) is independent of the differentiation of the larval motor neurons as the polarity switch in DD NMJs occurs normally in *lin-6* mutant animals in which the generation of postembryonic motor neurons is prevented (Hallam and Jin, 1998; White *et al.,* 1978). However, the elimination of motor neuron input to the DDs on the dorsal side could depend on inputs from ventral cholinergic neurons as DA and DB synapses onto the DDs are retained in *lin-6* mutant animals (White *et al.,* 1978).

III. Genetic Control of Motor Neuron Differentiation

As previously described, the lineal origin, morphology, and connectivity of the *C. elegans* motor neuron circuit have been defined at high resolution. The reproducibility of these characteristics from animal to animal is indicative of a strong program of genetic control. This idea is consistent with the isolation of a wide array of mutants that perturb specific motor neuron traits. Of particular interest are genes that encode transcription factors (Table II). These molecules are key determinants of neuronal identity through their capacity to regulate multiple downstream targets. Studies in both vertebrate and invertebrate systems have shown that transcriptional control of cell-specific gene expression is critically important at every stage of neuronal development from the early production of neuronal progenitor cells to the terminal cell divisions that generate individual motor neuron classes (Brody and Odenwald, 2002; Shirasaki and Pfaff, 2002). Similar roles at various stages of motor neuron differentiation have also been assigned to specific transcription factors in *C. elegans* (Hallam *et al.,* 2000; Jin *et al.,* 1994; Miller *et al.,* 1992; Prasad *et al.,* 1998). Given the precision with which the *C. elegans* motor neuron lineages and morphology are now known, detailed studies of these transcription factors and their targets have the potential of providing an extraordinarily detailed view of how these key traits are specified by the genome.

TABLE II
TRANSCRIPTION FACTORS THAT SPECIFY MOTOR NEURON FATE

		Cholinergic		GABAergic	
Gene	Protein	Embryonic	Larval	Embryonic	Larval
cnd-1	bHLH	DA, DB		DD	
pag-3	Zn finger/ Gfi-1		VA, VB, VC		
zag-1	δEF1/ZFH-1	All	All	All	All
unc-30	Pitx2 HD			DD	VD
unc-55	COUP		AS		VD
alr-1	ARX HD			DD, RMEL/R, AVL	VD, DVB
lim-6	LIM HD			RMEL/R, AVL	DVB
arh-1	aryl hyodrocarbon receptor (AHR)			RMEL/R	
aha-1	AHR nuclear translocator			RMEL/R	
ceh-10	CHX10 HD			RMED	
lim-4	LIM HD			RMEL	
unc-3	olf-1/EBF	DA, DB	VA, VB, AS		
unc-4	UNC-4 HD	DA (SAB)	VA, VC		
vab-7	Eve HD	DB	VC		
ceh-12	HB9 HD		VB		
lin-11	LIM HD		VC		

A. Transcription Factors That Specify Pan-Neural Motor Neuron Fates or Which Function in Motor Neuron Progenitors

Mutant analysis has revealed at least three transcription factors that function either at early stages of the motor neuron lineages or which specify a broad range of characteristics expressed in several different motor neuron classes.

B. CND-1 Specifies Embryonic Motor Neurons

cnd-1 encodes a bHLH transcription factor that is most closely related to the NeuroD/Neurogenin subfamilies (Hallam *et al.,* 2000). bHLH transcription factors are widely employed in both vertebrates and invertebrates to govern a series of cell fate decisions in lineages giving rise to specific types of neurons (Guillemot, 1999; Shirasaki and Pfaff, 2002). CND-1 is expressed early in embryonic development in neural progenitor cells, is maintained in postmitotic DA, DB, and DD motor neurons after hatching, but fades as post-embryonic ventral cord differentiation is initiated. Mutations in the *cnd-1* locus disrupt differentiation of embryonic DA, DB, and DD ventral cord motor classes; the normal complement of these motor neurons is reduced, class-specific markers are miss-expressed, and motor neuron axonal outgrowth and placement in the ventral cord are disrupted. These findings suggest that CND-1 may fulfill the dual roles of regulating gene expression in the embryonic motor neuron lineages as well as controlling class-specific traits in postmitotic cells (Hallam *et al.,* 2000). In its early role, CND-1 could delay exit from the cell cycle as has been shown for Neurogenins (Shirasaki and Pfaff, 2002). Later, CND-1 regulates expression of homeodomain proteins (e.g., *unc-3, unc-4, unc-30*; see the following text) that define the fates of postmitotic neurons (Hallam *et al.,* 2000). This finding parallels recent evidence showing that bHLH factors Olig2 and NGN2 are key regulators of motor neuron fate in the vertebrate spinal cord where they coordinate cell cycle exit with expression of homeodomain transcription factors that define motor neuron identity (Lee *et al.,* 2005). The evolutionary conservation of the bHLH function in the differentiation of the motor circuit predicts that other bHLH factors are likely to act in concert with CND-1 in the embryonic motor neuron lineage. The absence of CND-1 in the postembryonic motor neuron lineage suggests that additional bHLH factors may be required in this developmental stage. One candidate for this role, HLH-3, an *achaete-scute* homolog, is selectively expressed in the P-cell neuroblast and its daughter cells in L1 stage larvae (Fig. 5) (R. Doonan and A. Alfonso, 2004, personal communication; Krause *et al.,* 1997).

C. PAG-3 Controls Specific Post-Embryonic Motor Neuron Fates

pag-3 encodes a Zn-finger transcription factor similar to mammalian Gfi-1 and *Drosophila* Senseless proteins. As opposed to CND-1, which controls embryonic motor neuron differentiation, PAG-3 is exclusively required in the larval motor neuron lineage. A role in motor neuron fate was initially suggested by the strong locomotory defect observed in *pag-3* mutants (Jia *et al.,* 1996, 1997). Subsequent analysis showed that PAG-3 functions in the P-cell–derived sublineage giving rise to VA, VB, and VC motor neurons (Fig. 6). Normally, the Pn.aa cell generates two daughters, Pn.aap, which is fated to yield a VC motor neuron (or undergo programmed cell death) and Pn.aaa which divides again to produce VA and VB (Fig. 5) (Sulston and Horvitz, 1977). In *pag-3* mutants, Pn.aaa daughters variably reiterate the fate of Pn.aa mother cells to produce extra VC motor neurons (or cell deaths), a few VAs, and no detectable VB motor neurons as judged by expression of class-specific markers. Thus, PAG-3 function resembles that of UNC-86, which also distinguishes mother-daughter fates in other neuronal lineages. PAG-3 protein expression is sustained in postmitotic VA and VB motor neurons which suggests that it may also specify class specific traits in these cells (Cameron *et al.,* 2002).

D. ZAG-1 Defines Post-Mitotic Motor Neuron Traits

ZAG-1 and its homologs in flies (Zfh1) and vertebrates (δEF1) are large complex proteins containing both Zn finger and homeobox DNA-binding domains (Clark and Chiu, 2003; Wacker *et al.,* 2003). *zag-1* mutants in *C. elegans* display a wide range of neuronal defects including misexpression of neuron-specific markers, axon guidance errors, and nerve cord defasciculation. The uncoordinated phenotype that *zag-1* mutants display is correlated with transient expression of *zag-1*:GFP reporters in postmitotic motor neurons. In contrast to *cnd-1* and *pag-3* mutants, motor neuron numbers are normal in *zag-1* mutant animals. This result and the varied effects of *zag-1* on selected motor neuron traits suggest that ZAG-1 may control specific subsets of genes in each motor neuron class. In this model, ZAG-1 is likely to function downstream of CND-1 and PAG-3 (Clark and Chiu 2003; Wacker *et al.,* 2003). Zfh1 and δEF1 are also highly expressed in the fly and mouse nervous systems, respectively, but neuronal defects have not been described in the existing mutants for these genes (Broihier *et al.,* 1998). In other tissues, δEF1 functions as a transcriptional repressor and antagonizes bHLH activity by competing for a common DNA-binding site (Sekido *et al.,* 1994). It would be worthwhile to determine if ZAG-1 utilizes a similar strategy in the *C. elegans* motor neuron circuit and thereby potentially define a primordial mechanism for ZAG-1 protein action in the specification of neuronal fate.

E. Transcription Factors That Specify Motor Neuron Class and Neurotransmitter Phenotype

Ultimately, the postmitotic cells generated by each motor neuron lineage must express selected physiological traits and establish connections in the locomotory circuit that satisfy the functional requirements of the corresponding motor neuron class. Transcription factors that control these decisions have been identified largely from forward genetic screens for animals with defective motor neuron morphology or abnormal movement. With one exception (UNC-55; see the following text), these transcription factors segregate into two groups on the basis of the neurotransmitter, GABA, or acetylcholine, used by the motor neuron in which they function (Table II).

1. *Specification of GABAergic Motor Neurons*

There are 25 GABA motor neurons in *C. elegans*: 19 D-type (DD, VD), 4 RME, AVL, and DVB (McIntire *et al.*, 1993). As previously outlined, separate sets of transcription factors define the D-type motor neurons in the VNC versus RME, AVL, and DVB. (Table II and the following text).

a. Transcription Factors that Specify GABA Motor Neurons in the Ventral Cord. Mutations in the *unc-30* gene were initially recognized as animals showing a "shrinker" phenotype in which muscles on the dorsal and ventral sides contract simultaneously rather than alternately as in normal movement (Jin *et al.*, 1994). This behavioral defect is common to mutations that disrupt the contralateral inhibitory functions of DD and VD motor neurons (McIntire *et al.*, 1993a). *unc-30* encodes a Prd-type homeodomain transcription factor that is expressed in all 19 members of the DD and VD classes (Jin *et al.*, 1994). UNC-30 controls a suite of downstream targets highly expressed in these cells including two hallmarks of GABAergic identity, the GABA synthetic enzyme, glutamate decarboxylase (GAD) (UNC-25), and the GABA vesicular transporter (UNC-47) (Cinar *et al.*, 2005; Eastman *et al.*, 1999). DD and VD motor neurons also show axon guidance and fasciculation errors. Together, these mutant defects are indicative of a broad role for UNC-30 in regulating GABA motor neuron differentiation. The dependence of GABA synthesis on UNC-30 activity is evolutionarily conserved; Pitx2, the homolog of UNC-30, is expressed in mammalian GABA neurons, directly regulates transcription of the glutamate decarboxylase gene in the mouse, and can restore *unc-25* expression to an *unc-30* mutant worm (Westmoreland *et al.*, 2001).

In contrast to the role of UNC-30, which specifies shared DD and VD characteristics, UNC-55, a COUP domain steroid hormone receptor transcription factor, defines a key morphological trait that distinguishes these D-class motor neuron subtypes. Adult DD and VD motor neurons normally adopt opposite polarities; DD motor neurons synapse with dorsal muscles whereas VDs innervate ventral muscles (Fig. 4) (White *et al.*, 1986). In *unc-55* mutants, however, VD motor neurons undergo a "homeotic" transformation in which they now adopt the dorsal output polarity usually reserved for the DDs (Walthall, 1990). UNC-55 protein is selectively expressed in VD (and AS) motor neurons soon after their birth in the late L1 (Zhou and Walthall, 1998). Thus, UNC-55 may function in the VDs to block expression of genes that lead to the adoption of dorsal outputs (Shan *et al.*, 2005). This genetic program is presumptively activated in the DDs in the first larval stage to drive a developmentally timed switch from ventral to dorsal polarity. Synapses are removed at one end of the DD motor neuron and reassembled at the other in a remodeling event that does not involve the withdrawal of either dendritic or axonal processes (Hallam and Jin, 1998; White *et al.*, 1978). Support for the idea that UNC-55 acts in opposition to this program is provided by an elegant experiment in which ectopic expression of UNC-55 in DD motor neurons blocks the L1 polarity reversal and "freezes" the DDs in their ventral output configuration (Shan *et al.*, 2005). The proposed transcriptional repressor function of UNC-55 is consistent with the finding that *flp-13*, a neuropeptide encoding gene expressed in DD motor neurons, contains an upstream COUP domain binding site through which UNC-55 inhibits *flp-13* expression in the VDs (Shan *et al.*, 2005). UNC-55 appears to antagonize UNC-30 which binds an adjacent site to activate *flp-13* transcription. Although FLP-13 is unlikely to exercise a necessary role in the DD polarity reversal, it will be interesting to determine if other downstream genes that are required for this event are also activated by UNC-30 (Shan *et al.*, 2005). The identification of the subset of these genes that are repressed by UNC-55 should lead to a molecular understanding of this interesting developmental phenomenon and thereby shed light on fundamental processes that control neuronal plasticity by coordinating polarity with synaptic remodeling (Bradke and Dotti, 2000).

alr-1 encodes a conserved member of the Aristaless/ARX family of homeodomain proteins. Although ALR-1 is expressed in all GABA motor neurons, known roles for ALR-1 in GABA motor neuron differentiation are limited to the D-class motor neurons (Melkman and Sengupta, 2005). ALR-1 blocks expression of the DD marker, *flp-13*, in VD motor neuron and thus may function in parallel with UNC-55 to distinguish VD versus DD fates. VD motor neuron polarity is not perturbed in *alr-1* mutants, however, indicating that other UNC-55–regulated genes that define this VD-specific trait are independent of ALR-1 (Melkman and Sengupta, 2005). ALR-1 fulfills a separate role in the DD motor neurons in which ventral but not dorsal process morphology is abnormal in *alr-1* mutants (S. Barlow and D. M. Miller, 2004, unpublished data). These findings in *C. elegans* parallel observations that the vertebrate Aristaless homolog, ARX, is highly expressed in GABA neurons and that ARX mutations perturb GABA motor neuron differentiation in mammals (Kitamura *et al.*, 2002). Recent work has shown that similar defects arising from mutations in the human ARX gene may account for a significant proportion of known cases of X-linked mental retardation in males (Sherr, 2003; Stromme *et al.*, 2002). Thus, the discovery of a role for an ARX homolog in nematode GABA motor neuron differentiation offers a valuable opportunity to exploit this tractable model system to identify conserved ARX target genes with potentially important functions in the development of the human brain (Melkman and Sengupta, 2005).

2. *Transcription Factors That Specify GABA Motor Neurons in the Nerve Ring and in the Defecation Circuit*

The transcription factors, LIM-6 (LMX homeodomain), CEH-10 (CHX10 homeodomain), and ARH-1 (aryl hydrocarbon receptor) and its cofactor AHA-1 (ARH nuclear translocator) are expressed among the GABA motor neurons

RME, AVL, and DVB and thus complement UNC-30 expression in the D-class motor neurons (Forrester *et al.,* 1998; Hobert *et al.,* 1999; Huang *et al.,* 2004). As previously noted, UNC-30 controls key features of D-class motor differentiation including expression of UNC-25 (GAD) and UNC-47 (GABA vesicular transporter) (Eastman *et al.,* 1999). In contrast, different sets of transcription factors regulate these genes in the RME, AVL, and DVB motor neurons. For example, GABA expression is undetectable in RMEL in *ceh-10* mutants (Forrester *et al.,* 1998). In *lim-6* animals, *unc-25* expression is significantly reduced in AVL and DVB but *unc-47* expression in these cells is unaffected (Hobert *et al.,* 1999). Thus, the critically important functions of neurotransmitter synthesis and vesicular uptake are controlled by independent transcriptional mechanisms in different GABA motor neurons. Defective enteric muscle function in *lim-6* mutants is correlated with abnormal sprouting of AVL and DVB axons which suggests that both UNC-30 and LIM-6 control target genes that define GABA motor neuron morphology (Hobert *et al.,* 1999; Jin *et al.,* 1994).

A distinctive morphological difference between two of the four RME motor neurons, RMEL and RMER, is defined by AHR-1 and its cofactor AHA-1. Four RME neurons are arranged in symmetrical pairs around the nerve ring in the head, RMED (dorsal) versus RMEV (ventral) and RMEL (left) versus RMER (right). RMED/V are distinguished from RMED/L by the presence of single posteriorly directed process that joins either the DNC or VNC. In *ahr-1* or *aha-1* mutants, RMEL/R adopt the characteristic morphology of RMED/L. Conversely, ectopic expression of AHR-1 in RMED/V is sufficient to impose the morphological fate of RMEL/R. These findings indicate that *ahr-1* is selectively expressed in RMED/L to specify a subset of traits that distinguish these GABA motor neurons from other members of this class (Huang *et al.,* 2004).

3. *Specification of Cholinergic Motor Neurons*

Acetylcholine is the principle neurotransmitter utilized by excitatory motor neurons in nematodes and in the vertebrate spinal cord (Misgeld *et al.,* 2002; Rand and Nonet, 1997). Cholinergic motor neurons in these phyla express the synthetic enzyme, choline acetyltransferase (CHA-1) and acetylcholine vesicular transporter (UNC-17) (Alfonso *et al.,* 1993, 1994). In the *C. elegans* VNC, these cholinergic markers are detected in all A-type (DA, VA), B-type (DB, VB), and AS and VC motor neurons (Lickteig *et al.,* 2001; Rand and Nonet, 1997). Here we describe the roles of four transcription factors, UNC-3, UNC-4, VAB-7, and LIN-11 in the differentiation of these cells. Although a subset of motor neurons that innervate head muscles at the nerve ring are also cholinergic, transcription factors that specify these cell types have not been identified. Thus, the genetic control of cholinergic motor neuron differentiation parallels that of GABA motor neurons in which, as previously described, GABA motor neurons in the head and ventral cord are specified by separate sets of transcription factors.

F. UNC-3 Specifies Cholinergic Motor Neuron Fate

unc-3 encodes a highly conserved member of the O/E family of transcription factors showing ~80% identity within the Zn finger DNA binding and HLH dimerization domains (Dubois and Vincent, 2001; Prasad *et al.,* 1998). Strong UNC-3 expression is restricted to postmitotic cholinergic motor neurons (DA, VA, DB, VB, AS) in the ventral cord (Prasad *et al.,* 1998) (K. Lickteig and D. M. Miller, 2004, unpublished data) and to ASI chemosensory neurons in the head. In this respect, UNC-3 expression parallels that of its vertebrate O/E homologs in the motor column of the developing spinal cord and in the olfactory epithelium (Garel *et al.,* 1997; Wang *et al.,* 1997). Vertebrate O/E proteins are also expressed in other CNS and PNS neurons, in B lymphocytes, and in mesodermal cells which suggests that the developmental role for this protein family has expanded in evolution (Corradi *et al.,* 2003; Lin and Grosschedl, 1995; Mella *et al.,* 2004). *unc-3* mutants show a strong uncoordinated phenotype; the VNC is highly disorganized with misplaced and defasciculated axonal processes (Durbin 1987; Prasad *et al.,* 1998; Wightman *et al.,* 1997; J. White, 1997, personal communication). Mosaic analysis suggests that the movement defect is largely a consequence of UNC-3 dysfunction in motor neurons (Herman, 1987). This finding is consistent with the observation that UNC-3 is required for expression of CHA-1 and UNC-17 in ventral cord cholinergic motor neurons that also express UNC-3 (K. Lickteig, D. Frisby, J. Duerr, D. M. Miller, and J. Rand, 1997, unpublished data). UNC-3 may also control other class-specific traits within this group of cholinergic motor neurons (S. Cameron, personal communication; J. Meir and DMM, unpublished data). D-class motor neurons show axonal and synaptic defects in *unc-3* mutants although UNC-3 is presumptively not expressed in these cells (Prasad *et al.,* 1998; Yeh *et al.,* 2005). The results of some UNC-3 antibody staining experiments, however, indicate that UNC-3 may be transiently expressed in GABA motor neurons (Prasad *et al.,* 1998) and, therefore, potentially required for normal differentiation of all ventral cord motor neuron classes. Mutations in mammalian O/E proteins produce a wide array of neuronal defects including axon guidance errors and misexpression of neuronal markers (Garcia-Dominguez *et al.,* 2003; Garel *et al.,* 1999). The strong conservation of UNC-3 protein structure and function suggest that further studies in *C. elegans* should be useful for defining the fundamental roles of O/E proteins in neuronal differentiation.

G. The UNC-4 Homeodomain Protein Controls Synaptic Specificity in the Ventral Cord Motor Circuit

unc-4 mutants display a strong movement defect in which a touch to the head evokes a characteristic dorsal coiling response but no backward locomotion. Forward movement is also impaired but to a lesser extent. The biased effect of *unc-4*

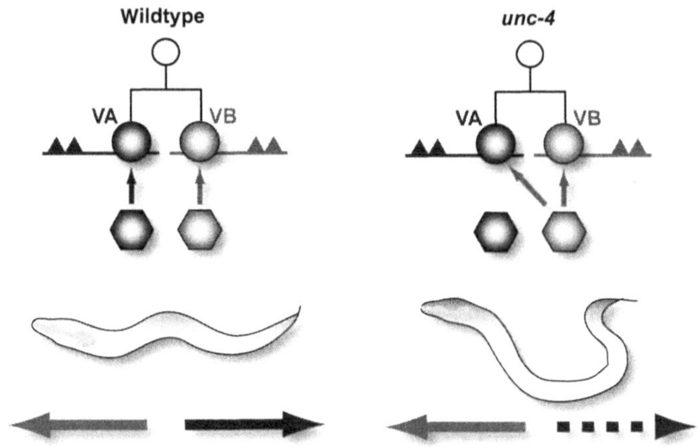

FIG. 7. VA motor neurons are miswired with VB-type inputs in *unc-4* mutants. Most VA and VB motor neurons arise from a common precursor and synapse with independent sets of command interneurons to control backward (blue) and forward (red) locomotion, respectively. Inputs to VA motor neurons are from AVA (gap junction, chemical synapse) and from AVD, AVE (chemical synapses). Command interneuron inputs to VBs are from AVB (gap junction) and PVC (chemical synapse). In *unc-4* mutants, VA motor neurons are miswired with inputs normally reserved for VB sisters. As a consequence, *unc-4* mutants are incapable of backward movement (dashed arrow). (See Color Insert.)

mutations on backward locomotion is correlated with a specific input defect to the VA class of excitatory motor neurons. EM reconstruction of the VNC from an *unc-4* mutant revealed that VA motor neurons are miswired with inputs normally reserved for sister VB motor neurons (White *et al.*, 1992). In *unc-4 (e120)*, the usual inputs to VAs from command interneurons AVA, AVD, and AVE are replaced with connections with AVB and PVC. These changes involve both chemical synapses and gap junctions (Fig. 7). Remarkably, Unc-4 synaptic defects are not accompanied by other alterations in the organization of the nerve cord. All of the affected neurons appear morphologically normal and occupy their usual locations in the ventral cord fascicle. Thus, UNC-4 appears to function as "neural specificity" locus, a gene that controls synaptic choice but not other differentiated traits of these neurons such as process outgrowth or fasciculation that might indirectly affect this decision (Miller and Niemeyer, 1995; White *et al.*, 1992).

H. UNC-4: Turning off the Wrong Genes to Make the Right Connections

Molecular analysis revealed that *unc-4* encodes the founding member of a family of Prd-like homeodomain transcription factors and that UNC-4 protein is highly expressed in VA motor neurons (Miller and Niemeyer, 1995; Miller *et al.*,

1992). These findings lead to the proposal that UNC-4 regulates specific target genes in the postsynaptic partner, the VA motor neurons, to prevent the adoption of VB-type inputs (Winnier *et al.,* 1999) (Fig. 7). UNC-4 could function, for example, as a transcriptional repressor to turn off VB genes that would otherwise mark the VAs for input from AVB and PVC. This idea was substantiated by two findings: (1) UNC-4 function depends on physical interaction with UNC-37/ Groucho, a highly conserved transcriptional co-repressor protein; and (2) VB motor neuron specific GFP reporters are ectopically expressed in VAs in *unc-4* and *unc-37* mutants (Miller *et al.,* 1993; Pflugrad *et al.,* 1997; Winnier *et al.,* 1999). The affected VB genes in this case, *acr-5* (nicotinic acetylcholine receptor), *del-1* (DEG/ENaC cation channel), and *glr-4* (non-NMDA type glutamate receptor) (Brockie *et al.,* 2001a) encode ion channels and presumptive cell surface components and, therefore, are attractive candidates for determinants of synaptic choice. However, deletion alleles of these genes fail to suppress the Unc-4 movement phenotype (i.e., restore backward locomotion) either as single mutants or in combination (S. E. Von Stetina and D. M. Miller, 2003, unpublished data). These results suggest either that these *unc-4* target genes (*acr-5, del-1, glr-4*) are not involved in synaptic choice or that their functions are redundantly provided by other as yet unknown downstream genes. We conclude that UNC-4 may control multiple downstream targets and that synaptic choice may depend on a limited subset of these genes. This model could also explain the failure to uncover Unc-4 suppressor mutations in genetic modifier screens designed to detect loss of function mutations in *unc-4* target genes (A. Winnier and D. M. Miller, 1998, unpublished data). The apparent insensitivity to this pathway to knock out mutations in single UNC-4 target genes has motivated the development of new cell-specific gene expression profiling strategies for identifying the full complement of UNC-4 regulated genes in VA motor neurons (Fox *et al.,* 2005; S. E. Von Stetina, R. Fox, D. M. Miller, 2005, unpublished data).

In addition to the VA motor neurons, UNC-4 is also expressed in 21 other neurons; I5 (1), SAB (3), DA (9), AVF (2), and VC (6) (Lickteig *et al.,* 2001; Miller and Niemeyer, 1995). Like the DAs, SAB motor neurons (in the retrovesicular ganglion at the anterior end of the ventral nerve cord) share key morphological and functional characteristics with VA motor neurons including cholinergic signaling, anteriorly directed axons, and synaptic inputs from AVA, AVD, and AVE (White *et al.,* 1986). The DA and SAB motor neurons are not miswired in *unc-4* mutants, however. Similarly, inputs to three of the VA motor neurons (VA1, VA11, VA12) are also unperturbed (White *et al.,* 1992). The affected VA motor neurons (VA2 to VA10) are distinguished by lineage as the only A-class motor neurons born with VB sister cells (Fig. 6B). Thus, UNC-4 may function in these VAs to repress a genetic program that is specifically activated in lineages that also give rise to VB motor neurons (White *et al.,* 1992). Perhaps UNC-4 function is not required in other A-class motor neurons for this purpose because genes leading to the creation of VA-type inputs are controlled by an independent pathway.

I. UNC-4 Controls Neurotransmitter Signaling Strength

In contrast to the role of UNC-4 in synaptic choice, which is limited to a subset of VA motor neurons, all of the motor neurons in which UNC-4 is normally expressed show a significant reduction in synaptic vesicles (SVs) in *unc-4* mutants (Fig. 8) (Lickteig *et al.*, 2001). For example, UNC-4 is expressed in VC motor neurons as they extend processes toward the vulva to innervate egg-laying muscles (Fig. 4). *unc-4* mutants display a specific egg-laying defect that may be a consequence of the observed reduction of SV proteins in

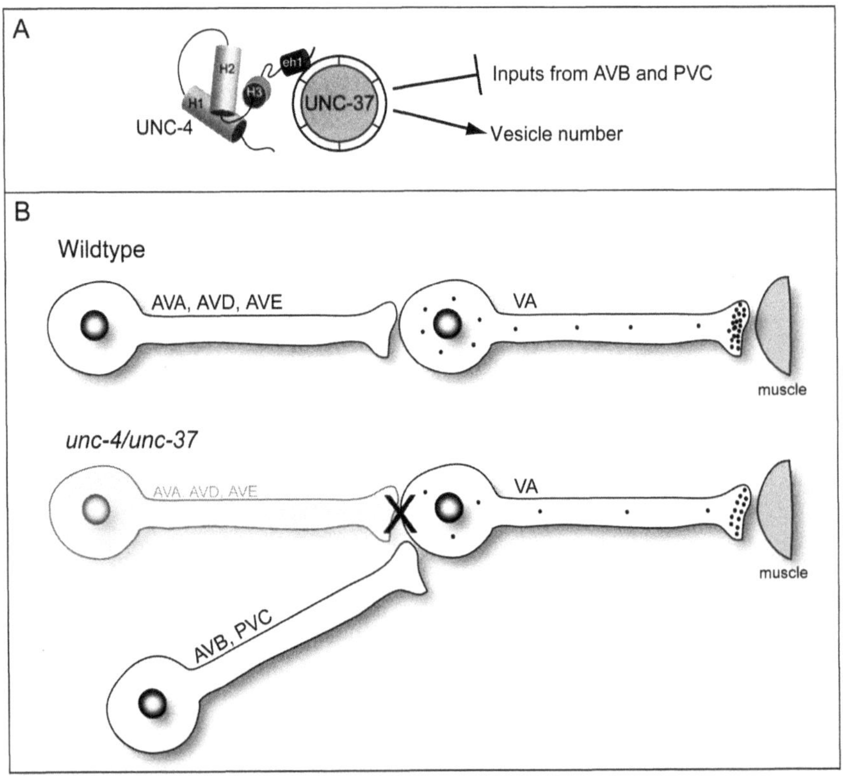

FIG. 8. Synaptic vesicles are reduced in *unc-4* mutant neurons. A. The UNC-4/UNC-37 complex maintains synaptic vesicle levels in all motor neurons that express UNC-4 and prevents the adoption of VB-type inputs by VA motor neurons. The UNC-4 protein includes a canonical DNA binding homeodomain comprised of 3 alpha helical regions (H1, H2, H3) and an eh1 sequence that interacts with UNC-37/Groucho. B. VA motor neurons are miswired in *unc-4* and *unc-37* mutants with inputs normally reserved for VB motor neurons (i.e. AVB, PVC) and show a 40% reduction in synaptic vesicle number.

these motor neurons (Bany *et al.*, 2003). It is important to point out that the *unc-4* mutation does not affect transcription of genes encoding SV proteins. Therefore, this phenotype is indirect and could involve negative effects on SV assembly or stability (Lickteig *et al.*, 2001). Because this reduction is not limited to the subset of VA motor neurons that are miswired in *unc-4* mutants, the UNC-4 target genes that control the strength of synaptic outputs may differ from the UNC-4 regulated genes that define the specificity of synaptic inputs (Fig. 8). The resolution of this question depends on the identification of these downstream genes.

J. VAB-7/Evenskipped Defines DB Motor Axon Polarity

vab-7 encodes the *C. elegans* homolog of the Even-skipped (Eve) homeodomain protein and functions both in embryonic patterning and motor neuron differentiation (Ahringer, 1996; Esmaeili *et al.*, 2002). *vab-7* mutants exhibit a distinctive forward Unc phenotype, suggesting B-type motor neuron dysfunction. VAB-7 protein is normally expressed in DB (but not VB) motor neurons. In the wild-type, DB motor neuron commissures turn posteriorly upon entering the dorsal nerve cord (Fig. 4). In *vab-7* mutants, however, DB axons reverse polarity and project toward the anterior (Fig. 9). In fact, this morphology is identical to that of DA motor neurons which normally extend anteriorly directed processes in the dorsal cord. The morphological similarity of the DB motor neurons to DA motor neurons in *vab-7* mutants suggests that VAB-7 functions to prevent DB motor neurons from adopting a DA fate. This model is supported by the observation that *unc-4*::GFP is ectopically expressed in DB motor neurons in *vab-7* mutant animals. Moreover, in *unc-4;vab-7* double mutants, normal posterior polarity is restored to DB motor neurons thereby indicating that ectopic UNC-4 expression is necessary for DBs to adopt DA-like anteriorly directed axons. A mutation in the *unc-37* (Groucho) locus also suppresses this axonal polarity defect which suggests that negative regulation of DB genes by the UNC-4/UNC-37 transcriptional repressor complex in *vab-7* mutants is sufficient to direct anterior axonal outgrowth. Consistent with this idea, the DB gene, *acr-5* (nAChR subunit), is turned off in *vab-7* mutants and then switched back on in *unc-4;vab-7* and in *unc-37;vab-7* double mutants (Esmaeili *et al.*, 2002). However, *acr-5* mutants display normal DB projections which rules out a necessary role for the ACR-5 acetylcholine receptor in directing DB axonal polarity (S. E. Von Stetina and D. M. Miller, 2000, unpublished results). Key receptors and guidance cues for steering axons along the D/V and L/R axes in *C. elegans* are now known. Axon guidance decisions along the A/P axis in *C. elegans*, however, are apparently determined by different mechanisms involving molecules that have yet to be identified. Thus, strategies that lead to the isolation of additional *vab-7* regulated

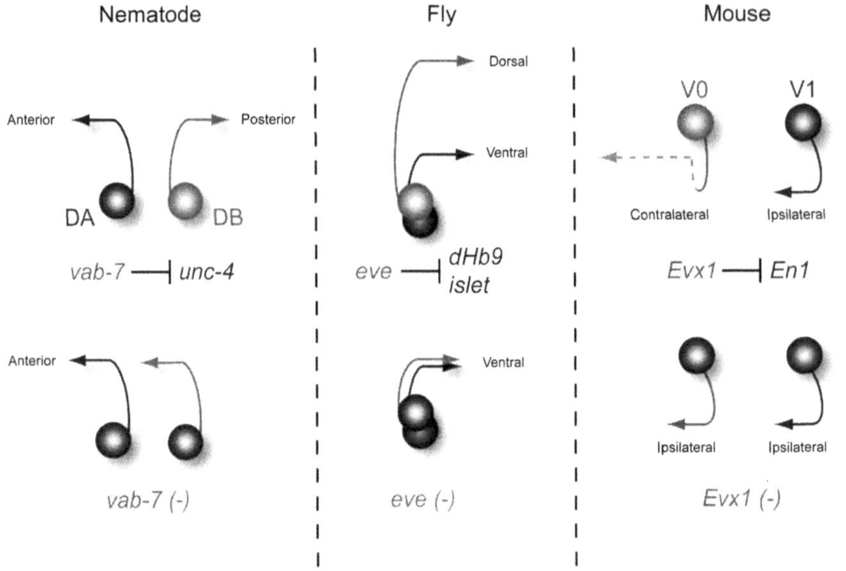

FIG. 9. Even-skipped (Eve) transcriptional repressor activity defines neuron identity and axonal trajectory in the axial motor circuits of diverse species. The *C. elegans eve* homolog, *vab-7*, negatively regulates *unc-4* to prevent DB motor neurons from adopting the anterior trajectory of DA motor neurons. In flies, dorsal motor neurons adopt a ventral trajectory when *islet* and *dHB9* are derepressed in *eve* mutants. The mouse *eve* homolog, *Evx1*, represses engrailed to maintain the medial location and contralateral axonal trajectory of V0 interneurons in the spinal cord. (See Color Insert.)

genes in DB motor neurons should identify strong candidates for these roles. It is interesting that Eve homologs in both flies and mice also control neuron identity and axonal polarity in the motor circuit (Fig. 9) (Landgraf *et al.*, 1999; Moran-Rivard *et al.*, 2001). In each case, Eve represses a target transcription factor that specifies an alternative neuron class. This striking conservation of Eve function from nematodes to mammals suggests that a detailed analysis of the VAB-7 pathway in *C. elegans* will elucidate a primordial mechanism of motor circuit neuron fate determination. For example, in *Drosophila,* Eve represses dHb9, a conserved homeodomain transcription factor expressed in motor neurons and in a subset of interneurons in flies and in the vertebrate spinal cord (Broihier and Skeath, 2002; Odden *et al.*, 2002; William *et al.*, 2003; Wilson *et al.*, 2005). A promoter GFP reporter for the *C. elegans* HB9 homolog, CEH-12, is exclusively expressed in VB class motor neurons (S. E. Von Stetina, R. Fox, and D. M. Miller, 2005, unpublished data). It will be interesting to determine if *ceh-12/Hb9* is also negatively regulated by VAB-7/Eve in *C. elegans*.

K. LIN-11/LIM1 Defines the Morphology of VC Motor Neurons

The *lin-11* locus, originally described for its cell division defects in the vulval lineage, is a founding member of the LIM family of homeodomain proteins (*L*in-11, *I*sl-1, *M*ec-3) (Freyd *et al.*, 1990). LIM homeodomain (HD) proteins are key determinants of motor neuron fate in flies and in the vertebrate spinal cord (Appel *et al.*, 1995; Shirasaki and Pfaff, 2002; Thor *et al.*, 1999). In *C. elegans*, *lin-11*::GFP expression in the VNC is limited to VC motor neurons (Hobert and Westphal, 2000; Hobert *et al.*, 1998). As opposed to other ventral cord motor neurons which provide input to muscle arms, the VC axons exit the cord in the mid body region to synapse with target vulval muscle soma in the egg laying circuit (Fig. 4). These axonal projections show defasciculation defects in *lin-11* mutants. The vertebrate LIN-11 homolog, LIM1, is also required for normal motor axon projections (Kania *et al.*, 2000). Axon guidance in this case depends on LIM1 regulation of the EphA receptor (Kania and Jessell, 2003). LIN-11 targets in *C. elegans* motor neurons have not been identified. Although the role for LIN-11 may parallel that of LIM1 in the specification of a motor neuron class, other *C. elegans* LIM-HDs have not been shown to affect motor neuron differentiation in the VNC. In flies and in vertebrates, motor neuron fate in the axial nerve cord is defined by unique patterns of overlapping LIM-HD expression (Hobert and Westphal, 2000). Perhaps this so-called "combinatorial code" of LIM-HD specification of motor neuron identity arose in evolution in concert with the increasing complexity of motor circuits.

IV. How Does It Work?

Muscle innervation in the trunk is bilaterally symmetrical and effectively limits body movements to undulations along the dorsal-ventral axis. *C. elegans* crawls on its side such that these waves of muscle contraction result in either forward or backward locomotion. Despite the simplicity of these locomotory behaviors, the neural basis of nematode movement is complex and is subject to modulation by a variety of different stimuli. In fact, movement can be viewed as the principle output of the *C. elegans* nervous system in which a key behavioral response efficiently integrates many different sensory inputs (Hobert, 2003). Here we will review current ideas about how the ventral cord motor circuit functions and also emphasize gaps in our knowledge that need to be addressed in order to build robust kinetic models of this network.

A. A Working Model of Nematode Locomotion

The wiring diagram provides a starting point for all working models of the ventral cord motor circuit (Fig. 10) (Chalfie and White, 1988; Driscoll and Kaplan, 1997; White *et al.*, 1976, 1986). Excitatory inputs to ventral muscles are provided by VA and VB motor neurons. Excitatory DA and DB motor neurons synapse with dorsal muscles. The D class GABAergic motor neurons, DD and VD, are proposed to function in a negative feedback loop in which muscle excitation on one side of the body triggers relaxation on the other. For example, VDs extend dendritic branches to join muscle arms at dyadic synapses with DA and DB excitatory motor neurons in the dorsal cord. In this circuit, excitation of dorsal muscles should also stimulate GABA release from the

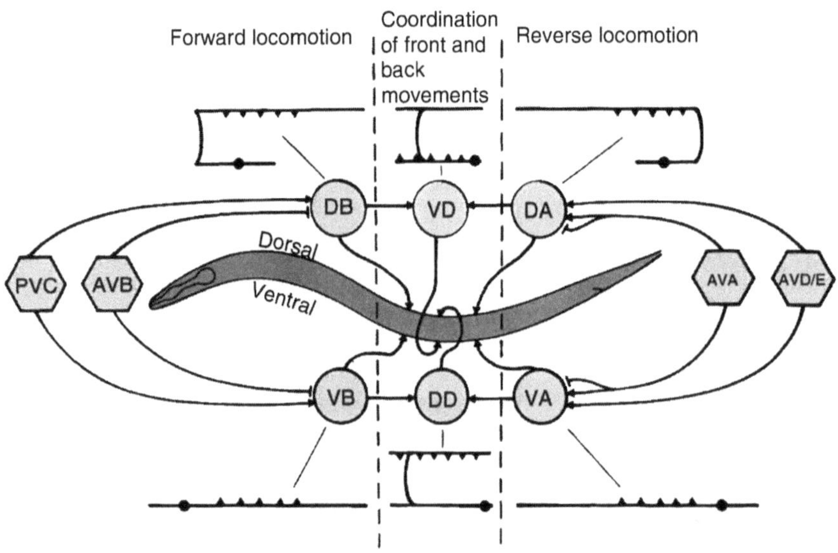

Fig. 10. Functional logic of the *C. elegans* motor circuit. Hexagons depict command interneurons (AVA, AVD, AVE) that synapse with excitatory motor neurons (DA, VA) involved in backward locomotion and command interneurons (AVB, PVC) that synapse with excitatory motor neurons (DB, VB) that drive forward movement. Arrows denote chemical synapses and blunt-ended lines are indicative of gap junctions (e.g., AVA makes chemical synapses and gap junctions with DA/VA). All inputs to the inhibitory motor neurons (DD, VD) are from other motor neurons: DD motor neurons innervate dorsal muscles and receive input from excitatory motor neurons (VA, VB) on the ventral side; VD motor neurons innervate ventral muscles and receive input from excitatory motor neurons (DA, DB) on the dorsal side. This arrangement provides a cross inhibitory circuit that coordinates muscle excitation on one side of the animal with relaxation on the other. Not shown are VC motor neurons, which have little impact on body movement, and AS motor neurons whose function has not been experimentally defined. (Adapted from Chalfie and White, 1988.)

corresponding VD axon on the ventral side to inhibit contralateral muscle contraction. Movement in either the forward or reverse directions depends on inputs from specific classes of command interneurons. AVA and AVD provide output to A class motor neurons (DA, VA) to drive backward locomotion (AVE provides similar inputs in the anterior ventral cord). Forward movement depends on AVB and PVC connections with B class motor neurons (DB, VB). In addition to linear information flow, interneurons through motor neurons to muscles, the motor circuit also enables much lateral signaling. Interneurons are connected to each other through chemical and electrical synapses. Motor neurons are connected by electrical junctions to neighboring motor neurons from the same class. And muscles in each row are also connected by electrical synapses to their neighbors.

B. Forward Versus Reverse Circuits

Laser ablation experiments have been used to assign specific interneurons and motor neurons to either the forward or reverse circuits. For example, removal of DA motor neurons selectively impairs backward locomotion whereas laser killing of DB motor neurons has the opposite effect of specifically perturbing forward movement. Although comparable laser ablation experiments have not been performed to directly establish roles for VA and VB motor neurons in the forward versus reverse circuits, these assignments are consistent with the shared morphologies and patterns of connectivity with the DAs and DBs, respectively (Chalfie *et al.,* 1985). This model is also supported by an experiment in which expression of a toxic MEC-4 (DEG/ENaC) cation channel subunit in A-class motor neurons blocks backward movement (Harbinder *et al.,* 1997).

Laser microsurgery has also been used to segregate the command interneurons into separate forward and reverse circuits. The removal of both AVA and AVD completely blocks backward movement whereas killing of AVB and PVC selectively impairs forward locomotion (Chalfie *et al.,* 1985; Wicks and Rankin, 1995). These results are consistent with the wiring diagram in which AVA and AVD synapse with A-class motor neurons (DA, VA) and AVB and PVC are connected to B-class motor neurons (DB, VB)(White *et al.,* 1986). This model is also supported by the Unc-4 mutant phenotype in which the inability to crawl backward is correlated with the miswiring of VA motor neurons with inputs from AVB and PVC (White *et al.,* 1992).

The decision to move either forward or backward is likely to depend on interactions between the command interneurons in each of these circuits. A simple model has been proposed (Chalfie *et al.,* 1985). During forward movement, activation of command interneurons in the "forward" circuit simultaneously inhibits backward locomotion. This inhibition can be overcome by specific sensory inputs. For example, tactile stimulation of anterior touch neurons is proposed to depolarize the backward command interneurons to drive the A-class motor neurons while simultaneously inhibiting the forward command interneurons. Extensive reciprocal connections between command interneurons in the forward and reverse circuits may account for these interactions. Additional experiments, however, have suggested that a mechanism for selecting forward versus reverse locomotion may be more widely distributed within the interneuron circuit. For example, hyperactivation of AVB by transgenic expression of a mutant glutamate receptor prolongs forward locomotion as expected but also extends the normal period of reverse locomotion. This finding suggests that AVB function may not be limited to the forward circuit (Zheng *et al.*, 1999).

C. How are Sinusoidal Waves Generated and Propagated?

The motor circuit previously described is used to generate waves of muscle contraction that propel the nematode forward or backward. This sinusoidal movement entails alternating dorsal/ventral excitation and relaxation of the body muscle. Although excitation of specific command interneurons is correlated with contractile waves moving in either the anterior (backward movement) or posterior (forward movement) directions, command interneuron excitability is prolonged and does not oscillate in synchrony with these movements (Brockie *et al.*, 2001b) (S. Lockery, personal communication). Even if sequential firing of ventral cord motor neurons were to depend on a propagating wave of command interneuron excitation, calculations based on the physiological measurements from the closely related nematode *Ascaris* and anatomical data from *C. elegans* suggest that the speed of electrical wave propagation is likely to be much faster than that of the muscle contraction wave (Niebur and Erdos, 1993). In this model, command interneurons are essentially equipotential throughout their length. Furthermore, although the command interneurons are likely to trigger rhythmic cycles in the motor circuit, they are not absolutely required for this activity as ablation of all of the command interneurons does not prevent movement (Zheng *et al.*, 1999). This result is analogous to the observation in vertebrate systems that the spinal cord includes intrinsic central pattern generators (CPGs) that can stimulate rhythmic muscle contractions in the absence of descending inputs from motor centers in the brain (Lanuza *et al.*, 2004).

The function of vertebrate CPGs depends on negative feedback loops that coordinate contraction and relaxation of opposing muscles. In these circuits, inhibitory signals are transmitted by GABAergic and glycinergic interneurons. As previously noted, the *C. elegans* GABA neurons can provide cross-inhibitory input to dorsal versus ventral muscles. Furthermore, depolarization induces slow activity oscillations in the inhibitory ventral cord motor neurons of *Ascaris* (Angstadt and Stretton, 1989). Although a role for this rhythmic activity in locomotion is plausible, direct measurements of *C. elegans* inhibitory motor neurons to detect this activity have not been performed. In addition, CPG function depends on cross-inhibition of excitatory motor neurons in the contralateral circuit. This feedback may exist in *Ascaris* as excitatory motor neurons are responsive to exogenous application of GABA (Davis and Stretton, 1996). GABA ionotropic and metabotropic receptors are highly enriched in microarray profiles of *C. elegans* excitatory motor neurons which suggests that these cells are at least potentially responsive to inhibitory signals (Fox *et al.,* 2005). In any case, these GABA motor neurons are surprisingly dispensable for movement as mutants defective for GABA signaling are in fact capable of movement. Only when prodded do the animals shrink due to simultaneous contraction of dorsal and ventral muscles (McIntire *et al.,* 1993a). Thus, it seems that during normal quiet movement other mechanisms are also operating.

The ventral cord excitatory motor neurons are also potential candidates for generators and propagators of the electrical activity wave. The processes of forward (DB and VB) and backward (VA and VB) excitatory motor neurons orient in opposite directions (Fig. 4). Thus, the direction of an electrical wave of excitation traveling through these processes is appropriately determined by their anatomy. Interestingly, in *vab-7* mutants orientation of DB motor neuron processes is reversed and forward movement is defective, supporting a causal relationship between process orientation and movement (Esmaeili *et al.,* 2002) (Fig. 9). White *et al.* (1986) also observed that the cholinergic motor neurons display elongated processes extending well past the active zone and proposed that these structures could serve as stretch receptors (White *et al.,* 1986). In this model, stretch activation of mechanosensitive channels in distal regions of a given excitatory motor neuron could potentiate neurotransmitter release and hence muscle contraction at the active zone (Tavernarakis *et al.,* 1997). Furthermore, overlapping stretch receptors from adjacent motor neurons of this class could also be affected by local bending and thereby propagate the wave of muscle contraction down the animal. The fundamental problem with this model is to imagine a mechanism whereby command interneurons trigger these oscillations. As previously noted, the predicted rapid transmission of electrical impulse within the command interneuron would be expected to excite all of the members of a given class of motor neurons (e.g., A or B) simultaneously (Niebur and Erdos, 1993). In addition, gap junctions between members of each motor neuron class could also propagate this electrical activity. Perhaps interneuron stimulation is subthreshold such that local stochastic oscillations, modulated by stretch, in motor neuron activity are required to trigger synaptic release at the neuromuscular junction. The electrical conductance of gap junctions between motor neurons is also unknown and, in this case, could be inefficient.

V. Conclusions and Future Directions

The simplicity and well-defined structure of the *C. elegans* motor circuit offer an unparalleled opportunity to define the genetic rules that govern the creation and function of this network. This point seems clear. A relevant question to ask, however, is whether this information from the nematode will also deepen our understanding of the much more complex and sophisticated motor circuits in the vertebrate spinal cord. This question can be answered in part by noting the strong evolutionary conservation of transcription factors and the mechanisms they employ to specify motor neuron identity. For example, molecular genetic analysis established that the UNC-4 homeodomain (HD) functions with UNC-37/Groucho as a negative regulator of gene expression in *C. elegans* motor neurons. This mechanism depends on an eh1 Groucho-interaction domain located in the UNC-4 C-terminal region (Miller *et al.,* 1993; Pflugrad *et al.,* 1997; Winnier *et al.,* 1999). Groucho proteins and homologs of UNC-4 that also contain a C-terminal eh1 domain are expressed in the developing spinal cord (Dehni *et al.,* 1995; Mansouri *et al.,* 1997). Moreover, subsequent studies have identified many additional classes of eh1-bearing HD transcription factors that also function with Groucho to specify neuronal fates in the vertebrate motor circuit (Muhr *et al.,* 2001). In these cases, specific neuronal progenitor domains are defined by cross-repressive interactions between different transcription factors (Shirasaki and Pfaff, 2002). A similar model applies to the specification of postmitotic motor neuron fates in *C. elegans.*

As noted earlier, UNC-4 and the Eve homolog, VAB-7, are expressed in adjacent DA and DB motor neurons, respectively. In *vab-7* mutants, UNC-4 is de-repressed to transform the prospective DB motor neuron into a DA type. The most dramatic evidence of this transformation is the adoption of the DA axonal trajectory by the *vab-7* mutant neuron (Esmaeili *et al.,* 2002). The VAB-7 homolog, Eve, also specifies motor neuron identity and axonal trajectory in flies (Landgraf *et al.,* 1999). The mammalian Eve homolog, Evx1, however, distinguishes fates of interneurons arising from adjacent V0 versus V1 domains (Moran-Rivard *et al.,* 2001). Thus, Eve defines motor neuron identity in nematodes and insects but is strictly required for interneuron fate in mammals (Fig. 9). One explanation for this difference is that Eve expressing motor neurons have been "lost" in the evolution of vertebrate motor circuits (Thor and Thomas, 2002). An alternative idea takes note of morphological and functional similarities between nematode motor neurons and vertebrate interneurons. In mammals, interneurons are confined to the spinal cord where they synapse with motor neurons. Similarly, *C. elegans* cholinergic motor neurons also establish connections within the nerve bundle. And, in addition to muscle arms, GABAergic motor neurons are also postsynaptic at these sites. These dyadic synapses provide an efficient structural mechanism for eliciting muscle contraction on one side of the animal while simultaneously inducing, through stimulation of GABA motor neuron activity, relaxation of body muscles on the opposite side (Driscoll and Kaplan, 1997; White *et al.,* 1976). Thus, the GABA motor neurons in this negative feedback loop are functionally analogous to inhibitory interneurons that modulate motor neuron activity in local circuits in the vertebrate spinal cord (Sharma and Peng, 2001; Xu *et al.,* 2005). Moreover, homologs of the transcription factors ALR-1 (ARX) and UNC-30 (Pitx2) that specify GABA neuron fate in *C. elegans* also regulate GABA interneuron gene expression and differentiation in the mammalian brain (Eastman *et al.,* 1999; Kitamura *et al.,* 2002; Melkman and Sengupta, 2005; Westmoreland *et al.,* 2001). Therefore, we favor a model in which spinal cord interneurons are evolutionary descendents of motor neurons in simple circuits like that of *C. elegans.*

This model also assumes that mammalian motor neurons may have evolved from specific invertebrate motor neurons. This idea is supported by the conserved function of the HB9 transcription factor in motor neuron differentiation. The HB9 homolog, CEH-12, is exclusively expressed in one class of *C. elegans* motor neurons (S. E. Von Stetina, R. Fox, and D. M. Miller, 2005, unpublished data). Similarly, in insects, dHb9 is detected in specific motor neuron types (Broihier and Skeath, 2002; Broihier *et al.,* 2004; Odden *et al.,* 2002). In birds and in mammals, HB9 expression broadens to include all classes of motor neurons in the developing spinal cord (Arber *et al.,* 1999; Thaler *et al.,* 1999). (HB9 is also expressed in a subset of interneurons in flies and vertebrates.) Genetic analysis has established that HB9 specifies motor neuron identity in nematodes, flies and mammals (S. E. Von Stetina, R. Fox, D. M. Miller, 2005, unpublished data). These findings strongly suggest that roles for Hb9 in the differentiation of axial motor circuits are evolutionarily conserved from worms to humans. At a molecular level, this conclusion must mean that HB9 target genes are also likely to be conserved in these pathways. Thus, we believe that experimental approaches that exploit the tractability of *C. elegans* to identify these transcriptional targets and their roles in motor neuron fate will also identify key effectors of motor circuit differentiation and function in the spinal cord.

However, additional fundamental information is needed in order to fully exploit this organism for these purposes. Our poor understanding of how the motor circuit actually drives locomotion is particularly glaring. Although a robust wiring diagram has been generated for this circuit, in most cases we do not know either the strength or sign of these connections. This problem can be rectified in part by the application of sophisticated electrophysiological analysis. This approach is still technically challenging in *C. elegans* but several labs have now successfully recorded from identified neurons (Brockie *et al.,* 2001; Goodman *et al.,* 1998; O'Hagan *et al.,* 2005). Non-invasive approaches such as *in vivo* imaging of calcium transients in key motor circuit neurons are now also possible

(Suzuki *et al.,* 2003) and will be especially important for monitoring activity during unfettered locomotion (S. Lockery, 2005, personal communication). The long-standing proposal that rhythmic motor neuron activity could depend on feedback from local stretch receptors is attractive but remains to be substantiated (Driscoll and Kaplan, 1997; Tavernarakis *et al.,* 1997; White *et al.,* 1986). Models of locomotion must also accommodate signals emanating from cells that may not be explicitly connected to the motor circuit in the wiring diagram. Genetic and physiological evidence indicates that motor neuron activity, for example, can be highly modulated through G-protein–coupled receptors to classical neurotransmitters (e.g., dopamine, serotonin) and to neuropeptides (Chase *et al.,* 2004; Davis and Stretton, 1996; Jacob and Kaplan, 2003; Miller *et al.,* 1999; Nurrish *et al.,* 1999). In addition, the repertoire of potential signaling molecules and receptors affecting motor neuron function has been substantially expanded by the application of new methods for generating cell-specific gene expression profiles (Cinar *et al.,* 2005; Fox *et al.,* 2005; McKay *et al.,* 2003; Touroutine *et al.,* 2005). Of particular interest is the discovery that specific nicotinic acetylcholine receptors (nAChRs) are highly expressed in excitatory motor neurons (Culetto *et al.,* 2004; Fox *et al.,* 2005). Although it seems likely that these potent ionotropic receptors should affect motor neuron activity, presynaptic command interneurons are reportedly not cholinergic (J. Duerr and Jim Rand, 2002, personal communication; Rand and Nonet, 1997) which, therefore, points to the possibility that excitatory motor neurons may be signaling each other or responding to autocrine ACh loops in the ventral cord circuit. In addition to defining mechanisms of motor circuit control of locomotion, *C. elegans* should also be especially useful for identifying pathways that control the assembly of this circuit. Synapses are highly stereotypical and, therefore, are likely to be largely defined by genetic mechanisms (Ackley and Jin, 2004). Optical methods for visualizing specific synapses in this circuit should facilitate powerful genetic screens for unraveling these pathways (Shen and Bargmann, 2003).

References

Ackley, B. D., and Jin, Y. (2004). Genetic analysis of synaptic target recognition and assembly. *Trends Neurosci.* **27,** 540–547.

Ahringer, J. (1996). Posterior patterning by the *Caenorhabditis elegans* even-skipped homolog vab-7. *Genes. Dev.* **10,** 1120–1130.

Alfonso, A., Grundahl, K., Duerr, J. S., Han, H. P., and Rand, J. B. (1993). The *Caenorhabditis elegans* unc-17 gene: A putative vesicular acetylcholine transporter. *Science* **261,** 617–619.

Alfonso, A., Grundahl, K., McManus, J. R., and Rand, J. B. (1994). Cloning and characterization of the choline acetyltransferase structural gene (cha-1) from *C. elegans*. *J. Neurosci.* **14,** 2290–2300.

Angstadt, J. D., and Stretton, A. O. (1989). Slow active potentials in ventral inhibitory motor neurons of the nematode Ascaris. *J. Comp. Physiol. [A]* **166,** 165–177.

Appel, B., and Eisen, J. S. (2003). Retinoids run rampant: Multiple roles during spinal cord and motor neuron development. *Neuron* **40,** 461–464.

Appel, B., Korzh, V., Glasgow, E., Thor, S., Edlund, T., Dawid, I. B., and Eisen, J. S. (1995). Motoneuron fate specification revealed by patterned LIM homeobox gene expression in embryonic zebrafish. *Development* **121,** 4117–4125.

Arber, S., Han, B., Mendelsohn, M., Smith, M., Jessell, T. M., and Sockanathan, S. (1999). Requirement for the homeobox gene Hb9 in the consolidation of motor neuron identity. *Neuron* **23,** 659–674.

Aurelio, O., Hall, D. H., and Hobert, O. (2002). Immunoglobulin-domain proteins required for maintenance of ventral nerve cord organization. *Science* **295,** 686–690.

Bany, I. A., Dong, M. Q., and Koelle, M. R. (2003). Genetic and cellular basis for acetylcholine inhibition of *Caenorhabditis elegans* egg-laying behavior. *J. Neurosci.* **23,** 8060–8069.

Bradke, F., and Dotti, C. G. (2000). Establishment of neuronal polarity: Lessons from cultured hippocampal neurons. *Curr. Opin. Neurobiol.* **10,** 574–581.

Brockie, P. J., Madsen, D. M., Zheng, Y., Mellem, J., and Maricq, A. V. (2001a). Differential expression of glutamate receptor subunits in the nervous system of *Caenorhabditis elegans* and their regulation by the homeodomain protein UNC-42. *J. Neurosci.* **21,** 1510–1522.

Brockie, P. J., Mellem, J. E., Hills, T., Madsen, D. M., and Maricq, A. V. (2001b). The *C. elegans* glutamate receptor subunit NMR-1 is required for slow NMDA-activated currents that regulate reversal frequency during locomotion. *Neuron* **31,** 617–630.

Brody, T., and Odenwald, W. F. (2002). Cellular diversity in the developing nervous system: A temporal view from *Drosophila. Development* **129,** 3763–3770.

Broihier, H. T., Kuzin, A., Zhu, Y., Odenwald, W., and Skeath, J. B. (2004). Drosophila homeodomain protein Nkx6 coordinates motoneuron subtype identity and axonogenesis. *Development* **131,** 5233–5242.

Broihier, H. T., Moore, L. A., Van Doren, M., Newman, S., and Lehmann, R. (1998). zfh-1 is required for germ cell migration and gonadal mesoderm development in *Drosophila. Development* **125,** 655–666.

Broihier, H. T., and Skeath, J. B. (2002). Drosophila homeodomain protein dHb9 directs neuronal fate via crossrepressive and cell-nonautonomous mechanisms. *Neuron* **35,** 39–50.

Cameron, S., Clark, S. G., McDermott, J. B., Aamodt, E., and Horvitz, H. R. (2002). PAG-3, a Zn-finger transcription factor, determines neuroblast fate in *C. elegans. Development* **129,** 1763–1774.

Chalfie, M., Sulston, J. E., White, J. G., Southgate, E., Thomson, J. N., and Brenner, S. (1985). The neural circuit for touch sensitivity in Caenorhabditis elegans. *J. Neurosci.* **5,** 956–964.

Chalfie, M., and White, J. (1988). The nervous system. *In* "The nematode *Caenorhabditis elegans*" (W. B. Wood, Ed.), pp. 337–391. Cold Spring Harbor Laboratory Press, Cold Spring Harbor, New York.

Chang, C., Yu, T. W., Bargmann, C. I., and Tessier-Lavigne, M. (2004). Inhibition of netrin-mediated axon attraction by a receptor protein tyrosine phosphatase. *Science* **305,** 103–106.

Chase, D. L., Pepper, J. S., and Koelle, M. R. (2004). Mechanism of extrasynaptic dopamine signaling in *Caenorhabditis elegans. Nat. Neurosci.* **7,** 1096–1103.

Christensen, M., Estevez, A., Yin, X., Fox, R., Morrison, R., McDonnell, M., Gleason, C., Miller, D. M. 3rd, and Strange, K. (2002). A primary culture system for functional analysis of *C. elegans* neurons and muscle cells. *Neuron* **33,** 503–514.

Cinar, H., Keles, S., and Jin, Y. (2005). Expression profiling of GABAergic motor neurons in *Caenorhabditis elegans. Curr. Biol.* **15,** 340–346.

Clark, S. G., and Chiu, C. (2003). *C. elegans* ZAG-1, a Zn-finger-homeodomain protein, regulates axonal development and neuronal differentiation. *Development* **130,** 3781–3794.

Colavita, A., Krishna, S., Zheng, H., Padgett, R. W., and Culotti, J. G. (1998). Pioneer axon guidance by UNC-129, a *C. elegans* TGF-beta. *Science* **281,** 706–709.

Corradi, A., Croci, L., Broccoli, V., Zecchini, S., Previtali, S., Wurst, W., Amadio, S., Maggi, R., Quattrini, A., and Consalez, G. G. (2003). Hypogonadotropic hypogonadism and peripheral neuropathy in Ebf2-null mice. *Development* **130,** 401–410.

Culetto, E., Baylis, H. A., Richmond, J. E., Jones, A. K., Fleming, J. T., Squire, M. D., Lewis, J. A., and Sattelle, D. B. (2004). The *Caenorhabditis elegans* unc-63 gene encodes a levamisole-sensitive nicotinic acetylcholine receptor alpha subunit. *J. Biol. Chem.* **279,** 42476–42783.

Dasen, J. S., Liu, J. P., and Jessell, T. M. (2003). Motor neuron columnar fate imposed by sequential phases of Hox-c activity. *Nature* **425,** 926–933.

Davis, R. E., and Stretton, A. O. (1996). The motornervous system of Ascaris: Electrophysiology and anatomy of the neurons and their control by neuromodulators. *Parasitology* **113,** (Suppl.), S97–117.

Dehni, G., Liu, T., Husain, J., and Stifani, S. (1995). TLE expression correlates with mouse embryonic segmentation, neurogenesis, and epithelial determination. *Mechan. Dev.* **53,** 369–381.

Dixon, S. J., and Roy, P. J. (2005). Muscle arm development in *Caenorhabditis elegans*. *Development* **132,** 3079–3092.

Driscoll, M., and Kaplan, J. (1997). Mechanotransduction. *In* "*C. elegans* II" (T. B. D. A. Riddle, B. J. Meyer, and J. R. Priess, Eds.), pp. 645–677. Cold Spring Harbor Press, Cold Spring Harbor, New York.

Dubois, L., and Vincent, A. (2001). The COE–Collier/Olf1/EBF–transcription factors: Structural conservation and diversity of developmental functions. *Mech. Dev.* **108,** 3–12.

Durbin, R. M. (1987). Studies on the development and organisation of the nervous system of *Caenorhabditis elegans*. University of Cambridge, England, Ph.D. dissertation.

Eastman, C., Horvitz, H. R., and Jin, Y. (1999). Coordinated transcriptional regulation of the unc-25 glutamic acid decarboxylase and the unc-47 GABA vesicular transporter by the Caenorhabditis elegans UNC-30 homeodomain protein. *J. Neurosci.* **19,** 6225–6234.

Esmaeili, B., Ross, J. M., Neades, C., Miller, D. M. 3rd, and Ahringer, J. (2002). The *C. elegans* even-skipped homologue, vab-7, specifies DB motoneurone identity and axon trajectory. *Development* **129,** 853–862.

Forrester, W. C., Perens, E., Zallen, J. A., and Garriga, G. (1998). Identification of Caenorhabditis elegans genes required for neuronal differentiation and migration. *Genetics* **148,** 151–165.

Fox, R. M., Von Stetina, S. E., Barlow, S. J., Shaffer, C., Olszewski, K. L., Moore, J. H., Dupuy, D., Vidal, M., and Miller, D. M. 3rd. (2005). A gene expression fingerprint of *C. elegans* embryonic motor neurons. *BMC Genomics* **6,** 42.

Freyd, G., Kim, S. K., and Horvitz, H. R. (1990). Novel cystein-rich motif and homeodomain in the product of the Caenorhabditis elegans cell lineage gene lin-11. *Nature* **344,** 876–879.

Garcia-Dominguez, M., Poquet, C., Garel, S., and Charnay, P. (2003). Ebf gene function is required for coupling neuronal differentiation and cell cycle exit. *Development* **130,** 6013–6025.

Garel, S., Marin, F., Grosschedl, R., and Charnay, P. (1999). Ebf1 controls early cell differentiation in the embryonic striatum. *Development* **126,** 5285–5294.

Garel, S., Marin, F., Mattei, M. G., Vesque, C., Vincent, A., and Charnay, P. (1997). Family of Ebf/Olf-1-related genes potentially involved in neuronal differentiation and regional specification in the central nervous system. *Dev. Dyn.* **210,** 191–205.

Goodman, M. B., Hall, D. H., Avery, L., and Lockery, S. R. (1998). Active currents regulate sensitivity and dynamic range in *C. elegans* neurons. *Neuron* **20,** 763–772.

Guillemot, F. (1999). Vertebrate bHLH genes and the determination of neuronal fates. *Exp. Cell. Res.* **253,** 357–364.

Hallam, S., Singer, E., Waring, D., and Jin, Y. (2000). The *C. elegans* NeuroD homolog cnd-1 functions in multiple aspects of motor neuron fate specification. *Development* **127,** 4239–4252.

Hallam, S. J., and Jin, Y. (1998). lin-14 regulates the timing of synaptic remodelling in *Caenorhabditis elegans*. *Nature* **395,** 78–82.

Harbinder, S., Tavernarakis, N., Herndon, L. A., Kinnell, M., Xu, S. Q., Fire, A., and Driscoll, M. (1997). Genetically targeted cell disruption in Caenorhabditis elegans. *Proc. Natl. Acad. Sci. USA* **94,** 13128–13133.

Hedgecock, E. M., Culotti, J. G., and Hall, D. H. (1990). The unc-5, unc-6, and unc-40, guide circumferential migrations of pioneer axons and mesodermal cells on the nematode epidermis. *Neuron* **2,** 61–85.

Herman, R. K. (1987). Mosaic analysis of two genes that affect nervous system structure in *Caenorhabditis elegans*. *Genetics* **116,** 377–388.

Hobert, O. (2003). Behavioral plasticity in *C. elegans*: Paradigms, circuits, genes. *J. Neurobiol.* **54,** 203–223.

Hobert, O., D'Alberti, T., Liu, Y., and Ruvkun, G. (1998). Control of neural development and function in a thermoregulatory network by the LIM homeobox gene lin-11. *J. Neurosci.* **18,** 2084–2096.

Hobert, O., Tessmar, K., and Ruvkun, G. (1999). The *Caenorhabditis elegans* lim-6 LIM homeobox gene regulates neurite outgrowth and function of particular GABAergic neurons. *Development* **126,** 1547–1562.

Hobert, O., and Westphal, H. (2000). Functions of LIM-homeobox genes. *TIG* **16,** 75–83.

Huang, X., Powell-Coffman, J. A., and Jin, Y. (2004). The AHR-1 aryl hydrocarbon receptor and its co-factor the AHA-1 aryl hydrocarbon receptor nuclear translocator specify GABAergic neuron cell fate in *C. elegans*. *Development* **131,** 819–828.

Hutter, H. (2003). Extracellular cues and pioneers act together to guide axons in the ventral cord of *C. elegans*. *Development* **130,** 5307–5318.

Jacob, T. C., and Kaplan, J. M. (2003). The EGL-21 carboxypeptidase E facilitates acetylcholine release at *Caenorhabditis elegans* neuromuscular junctions. *J. Neurosci.* **23,** 2122–2130.

Jia, Y., Xie, G., and Aamodt, E. (1996). *pag-3*, a *Caenorhabditis elegans* gene involved in touch neuron gene expression and coordinated movement. *Genetics* **142,** 141–147.

Jia, Y., Xie, G., McDermott, J. B., and Aamodt, E. (1997). The *C. elegans* gene pag-3 is homologous to the zinc finger proto-onogene gfi-1. *Development* **124,** 2063–2073.

Jin, Y., Hoskins, R., and Horvitz, H. R. (1994). Control of type-D GABAergic neuron differentiation by *C. elegans* UNC-30 homeodomain protein. *Nature* **372,** 780–783.

Kania, A., and Jessell, T. M. (2003). Topographic motor projections in the limb imposed by LIM homeodomain protein regulation of ephrin-A:EphA interactions. *Neuron* **38,** 581–596.

Kania, A., Johnson, R. L., and Jessell, T. M. (2000). Coordinate roles for LIM homeobox genes in directing the dorsoventral trajectory of motor axons in the vertebrate limb. *Cell* **102,** 161–173.

Kitamura, K., Yanazawa, M., Sugiyama, N., Miura, H., Iizuka-Kogo, A., Kusaka, M., Omichi, K., Suzuki, R., Kato-Fukui, Y., Kamiirisa, K., Matsuo, M., Kamijo, S., Kasahara, M., Yoshioka, H., Ogata, T., Fukuda, T., Kondo, I., Kato, M., Dobyns, W. B., Yokoyama, M., and Morohashi, K. (2002). Mutation of ARX causes abnormal development of forebrain and testes in mice and X-linked lissencephaly with abnormal genitalia in humans. *Nat. Genet.* **32,** 359–369.

Krause, M., Park, M., Zhang, J. M., Yuan, J., Harfe, B., Xu, S. Q., Greenwald, I., Cole, M., Paterson, B., and Fire, A. (1997). A *C. elegans* E/Daughterless bHLH protein marks neuronal but not striated muscle development. *Development* **124,** 2179–2189.

Landgraf, M., Roy, S., Prokop, A., VijayRaghavan, K., and Bate, M. (1999). even-skipped determines the dorsal growth of motor axons in *Drosophila*. *Neuron* **22,** 43–52.

Lanuza, G. M., Gosgnach, S., Pierani, A., Jessell, T. M., and Goulding, M. (2004). Genetic identification of spinal interneurons that coordinate left-right locomotor activity necessary for walking movements. *Neuron* **42,** 375–386.

Lee, S. K., Lee, B., Ruiz, E. C., and Pfaff, S. L. (2005). Olig2 and Ngn2 function in opposition to modulate gene expression in motor neuron progenitor cells. *Genes. Dev.* **19,** 282–294.

Lickteig, K. M., Duerr, J. S., Frisby, D. L., Hall, D. H., Rand, J. B., and Miller, D. M. 3rd (2001). Regulation of neurotransmitter vesicles by the homeodomain protein UNC-4 and its transcriptional corepressor UNC-37/groucho in *Caenorhabditis elegans* cholinergic motor neurons. *J. Neurosci.* **21,** 2001–2014.

Lin, H., and Grosschedl, R. (1995). Failure of B-cell differentiation in mice lacking the transcription factor EBF. *Nature* **376,** 263–267.

Liu, J. P., Laufer, E., and Jessell, T. M. (2001). Assigning the positional identity of spinal motor neurons: Rostrocaudal patterning of Hox-c expression by FGFs, Gdf11, and retinoids. *Neuron* **32,** 997–1012.

Mansouri, A., Yokota, Y., Wehr, R., Copeland, N. G., Jenkins, N. A., and Gruss, P. (1997). Paired-related murine homeobox gene expressed in the developing sclerotome, kidney, and nervous system. *Dev. Dyn.* **210,** 53–65.

McIntire, S. L., Jorgensen, E., and Horvitz, H. R. (1993a). Genes required for GABA function in *Caenorhabditis elegans. Nature* **364,** 334–337.

McIntire, S. L., Jorgensen, E., Kaplan, J., and Horvitz, H. R. (1993b). The GABAergic nervous system of *Caenorhabditis elegans. Nature* **364,** 337–341.

McKay, S. J., Johnsen, R., Khattra, J., Asano, J., Baillie, D. L., Chan, S., Dube, N., Fang, L., Goszczynski, B., Ha, E., Halfnight, E., Hollebakken, R., Huang, P., Hung, K., Jensen, V., Jones, J., S., H., Kai, Li, D., Mah, A., Marra, M., McGhee, J., Newbury, R., Pouzyrev, A., Riddle, D. L., Sonnhammer, E., Tian, H., Tu, D., Tyson, J. R., Vatcher, G., Warner, A., Wong, K., Zhao, Z., and Moerman, D. G. (2003). Gene expression profiling of cells, tissues, and developmental stages of the nematode *C. elegans. Cold Spring Harb. Symp. Quant. Biol.* **68,** 159–169.

Melkman, T., and Sengupta, P. (2005). Regulation of chemosensory and GABAergic motor neuron development by the *C. elegans* Aristaless/Arx homolog alr-1. *Development* **132,** 1935–1949.

Mella, S., Soula, C., Morello, D., Crozatier, M., and Vincent, A. (2004). Expression patterns of the coe/ebf transcription factor genes during chicken and mouse limb development. *Gene. Expr. Patterns* **4,** 537–542.

Miller, D. M. III, and Niemeyer, C. J. (1995). Expression of the *unc-4* homeoprotein in *Caenorhabditis elegans* motor neurons specifies presynaptic input. *Development* **121,** 2877–2866.

Miller, D. M. III, Niemeyer, C. J., and Chitkara, P. (1993). Dominant *unc-37* mutations suppress the movement defect of a homeodomain mutation in *unc-4*, a neural specificity gene in *Caenorhabditis elegans. Genetics* **135,** 741–753.

Miller, D. M., Shen, M. M., Shamu, C. E., Burglin, T. R., Ruvkun, G., Dubois, M. L., Ghee, M., and Wilson, L. (1992). *C. elegans* unc-4 gene encodes a homeodomain protein that determines the pattern of synaptic input to specific motor neurons. *Nature* **355,** 841–845.

Miller, K. G., Emerson, M. D., and Rand, J. B. (1999). Goalpha and diacylglycerol kinase negatively regulate the Gqalpha pathway in *C. elegans. Neuron* **24,** 323–333.

Misgeld, T., Burgess, R. W., Lewis, R. M., Cunningham, J. M., Lichtman, J. W., and Sanes, J. R. (2002). Roles of neurotransmitter in synapse formation: Development of neuromuscular junctions lacking choline acetyltransferase. *Neuron* **36,** 635–648.

Moerman, D. G., and Fire, A. (1997). Muscle. *In* "*C. elegans* II" (T. B. D. A. Riddle, B. J. Meyer, and J. R. Priess, Eds.), pp. 417–470. Cold Spring Harbor, New York.

Moran-Rivard, L., Kagawa, T., Saueressig, H., Gross, M. K., Burrill, J., and Goulding, M. (2001). Evx1 is a postmitotic determinant of v0 interneuron identity in the spinal cord. *Neuron* **29,** 385–399.

Muhr, J., Andersson, E., Persson, M., Jessell, T. M., and Ericson, J. (2001). Groucho-mediated transcriptional repression establishes progenitor cell pattern and neuronal fate in the ventral neural tube. *Cell* **104,** 861–873.

Niebur, E., and Erdos, P. (1993). Theory of the locomotion of nematodes: Control of the somatic motor neurons by interneurons. *Math. Biosci.* **118,** 51–82.

Nurrish, S., Segalat, L., and Kaplan, J. M. (1999). Serotonin inhibition of synaptic transmission: Galpha(0) decreases the abundance of UNC-13 at release sites. *Neuron* **24,** 231–242.

O'Hagan, R., Chalfie, M., and Goodman, M. B. (2005). The MEC-4 DEG/ENaC channel of *Caenorhabditis elegans* touch receptor neurons transduces mechanical signals. *Nat. Neurosci.* **8,** 43–50.

Odden, J. P., Holbrook, S., and Doe, C. Q. (2002). Drosophila HB9 is expressed in a subset of motoneurons and interneurons, where it regulates gene expression and axon pathfinding. *J. Neurosci.* **22,** 9143–9149.

Pflugrad, A., Meir, J. Y., Barnes, T. M., and Miller, D. M. III. (1997). The Groucho-like transcription factor UNC-37 functions with the neural specificity gene unc-4 to govern motor neuron identity in *C. elegans. Development* **124,** 1699–1709.

Plunkett, J. A., Simmons, R. B., and Walthall, W. W. (1996). Dynamic interactions between nerve and muscle in *Caenorhabditis elegans. Dev. Biol.* **175,** 154–165.

Prasad, B. C., Ye, B., Zackhary, R., Schrader, K., Seydoux, G., and Reed, R. R. (1998). *unc-3,* a gene required for axonal guidance in *Caenorhabditis elegans,* encodes a member of the O/E family of transcription factors. *Development* **125,** 1561–1568.

Rand, J., and Nonet, M. (1997). Synaptic Transmission. *In* "*C. elegans* II" (T. B. D. A. Riddle, B. J. Meyer, and J. R. Priess, Eds.), pp. 611–643. Cold Spring Harbor Press, Cold Spring Harbor, New York.

Ritzenthaler, S., and Chiba, A. (2003). Myopodia (postsynaptic filopodia) participate in synaptic target recognition. *J. Neurobiol.* **55,** 31–40.

Salser, S. J., Loer, C. M., and Kenyon, C. (1993). Multiple HOM-C gene interactions specify cell fates in the nematode central nervous system. *Genes Dev.* **7,** 1714–1724.

Sanes, J. R., and Lichtman, J. W. (1999). Development of the vertebrate neuromuscular junction. *Annu. Rev. Neurosci.* **22,** 389–442.

Sekido, R., Murai, K., Funahashi, J., Kamachi, Y., Fujisawa-Sehara, A., Nabeshima, Y., and Kondoh, H. (1994). The delta-crystallin enhancer-binding protein delta EF1 is a repressor of E2-box-mediated gene activation. *Mol. Cell. Biol.* **14,** 5692–5700.

Shan, G., Kim, K., Li, C., and Walthall, W. W. (2005). Convergent genetic programs regulate similarities and differences between related motor neuron classes in *Caenorhabditis elegans. Dev. Biol.* **280,** 494–503.

Sharma, K., and Peng, C. Y. (2001). Spinal motor circuits: Merging development and function. *Neuron* **29,** 321–324.

Shen, K., and Bargmann, C. I. (2003). The immunoglobulin superfamily protein SYG-1 determines the location of specific synapses in *C. elegans. Cell* **112,** 619–630.

Sherr, E. H. (2003). The ARX story (epilepsy, mental retardation, autism, and cerebral malformations): One gene leads to many phenotypes. *Curr. Opin. Pediatr.* **15,** 567–571.

Shirasaki, R., and Pfaff, S. L. (2002). Transcriptional codes and the control of neuronal identity. *Annu. Rev. Neurosci.* **25,** 251–281.

Starr, D. A., Hermann, G. J., Malone, C. J., Fixsen, W., Priess, J. R., Horvitz, H. R., and Han, M. (2001). unc-83 encodes a novel component of the nuclear envelope and is essential for proper nuclear migration. *Development* **128,** 5039–5050.

Stromme, P., Mangelsdorf, M. E., Shaw, M. A., Lower, K. M., Lewis, S. M., Bruyere, H., Lutcherath, V., Gedeon, A. K., Wallace, R. H., Scheffer, I. E., Turner, G., Partington, M., Frints, G., S., J. P., Fryns, Sutherland, G. R., Mulley, J. C., and Gecz, J. (2002). Mutations in the human ortholog of Aristaless cause X-linked mental retardation and epilepsy. *Nat. Genet.* **30,** 441–445.

Sulston, J. E. (1976). Post-embryonic development in the ventral cord of *Caenorhabditis elegans*. *Philos. Trans. R. Soc. Lond. B. Biol. Sci.* **275,** 287–297.

Sulston, J. E., and Horvitz, H. R. (1977). Post-embryonic cell lineages of the nematode, *Caenorhabditis elegans*. *Dev. Biol.* **56,** 110–156.

Sulston, J. E., Schierenberg, E., White, J. G., and Thomson, J. N. (1983). The embryonic cell lineage of the nematode *Caenorhabditis elegans*. *Dev. Biol.* **100,** 64–119.

Sulston, J. E., and White, J. G. (1980). Regulation and cell autonomy during postembryonic development of *Caenorhabditis elegans*. *Dev. Biol.* **78,** 577–597.

Suzuki, H., Kerr, R., Bianchi, L., Frokjaer-Jensen, C., Slone, D., Xue, J., Gerstbrein, B., Driscoll, M., and Schafer, W. R. (2003). *In vivo* imaging of *C. elegans* mechanosensory neurons demonstrates a specific role for the MEC-4 channel in the process of gentle touch sensation. *Neuron* **39,** 1005–1017.

Tavernarakis, N., Shreffler, W., Wang, S., and Driscoll, M. (1997). *unc-8*, a DEG/ENaC family member, encodes a subunit of a candidate mechanically gated channel that modulates *C. elegans* locomotion. *Neuron* **18,** 107–119.

Thaler, J., Harrison, K., Sharma, K., Lettieri, K., Kehrl, J., and Pfaff, S. L. (1999). Active suppression of interneuron programs within developing motor neurons revealed by analysis of homeodomain factor HB9. *Neuron* **23,** 675–687.

Thor, S., Andersson, S. G., Tomlinson, A., and Thomas, J. B. (1999). A LIM-homeodomain combinatorial code for motor-neuron pathway selection. *Nature* **397,** 76–80.

Thor, S., and Thomas, J. (2002). Motor neuron specification in worms, flies and mice: Conserved and 'lost' mechanisms. *Curr. Opin. Genet. Dev.* **12,** 558–564.

Touroutine, D., Fox, R. M., Von Stetina, S. E., Burdina, A., Miller, D. M. 3rd, and Richmond, J. E. (2005). ACR-16 encodes an essential subunit of the levamisole-resistant nicotinic receptor at the Caenorhabditis elegans neuromuscular junction. *J. Biol. Chem.* **280,** 27013–27021.

Wacker, I., Schwarz, V., Hedgecock, E. M., and Hutter, H. (2003). zag-1, a Zn-finger homeodomain transcription factor controlling neuronal differentiation and axon outgrowth in *C. elegans*. *Development* **130,** 3795–3805.

Wadsworth, W. G., and Hedgecock, E. M. (1992). Guidance of neuroblast migrations and axonal projections in *Caenorhabditis elegans*. *Curr. Opin. Neurobiol.* **2,** 36–41.

Walthall, W. W. (1990). Metamorphic-like changes in the nervous system of the nematode *Caenorhabditis elegans*. *J. Neurobiol.* **21,** 1085–1091.

Wang, S. S., Tsai, R. Y.,L., and Reed, R. R. (1997). The characterization of the Olf-1/EBF-like HLH transcription factor family: Implications in olfactory gene regulation and neuronal development. *J. Neurosci.* **17,** 4149–4158.

Westmoreland, J. J., McEwen, J., Moore, B. A., Jin, Y., and Condie, B. G. (2001). Conserved function of *Caenorhabditis elegans* UNC-30 and mouse Pitx2 in controlling GABAergic neuron differentiation. *J. Neurosci.* **21,** 6810–6819.

White, J. G., Albertson, D. G., and Anness, M. A. (1978). Connectivity changes in a class of motoneurone during the development of a nematode. *Nature* **271,** 764–766.

White, J. G., Albertson, D. G., and Anness, M. A. R. (1978). Connectivity changes in a class of motoneurone during the development of a nematode. *Nature* **271,** 764–766.

White, J. G., Southgate, E., and Thomson, J. N. (1992). Mutations in the *Caenorhabditis elegans unc-4* gene alter the synaptic input to ventral cord motor neurons. *Nature* **355,** 838–841.

White, J. G., Southgate, E., Thomson, J. N., and Brenner, S. (1976). Structure of the ventral nerve cord of *Caenorhabditis elegans*. *Phil. Trans. R. Soc. Lond.* **B275,** 327–348.

White, J. G., Southgate, E., Thomson, J. N., and Brenner, S. (1986). The structure of the nervous system of the nematode *Caenorhabditis elegans*. *Phil. Trans. R. Soc. Lond.* **B314,** 331–340.

Wicks, S. R., and Rankin, C. H. (1995). Integration of mechanosensory stimuli in *Caenorhabditis elegans*. *J. Neurosci.* **15,** 2434–2444.

Wightman, B., Baran, R., and Garriga, G. (1997). Genes that guide growth cones along the *C. elegans* ventral nerve cord. *Development* **124,** 2571–2580.

William, C. M., Tanabe, Y., and Jessell, T. M. (2003). Regulation of motor neuron subtype identity by repressor activity of Mnx class homeodomain proteins. *Development* **130,** 1523–1536.

Wilson, J. M., Hartley, R., Maxwell, D. J., Todd, A. J., Lieberam, I., Kaltschmidt, J. A., Yoshida, Y., Jessell, T. M., and Brownstone, R. M. (2005). Conditional rhythmicity of ventral spinal interneurons defined by expression of the Hb9 homeodomain protein. *J. Neurosci.* **25,** 5710–5719.

Winnier, A. R., Meir, J. Y., Ross, J. M., Tavernarakis, N., Driscoll, M., Ishihara, T., Katsura, I., and Miller, D. M. 3rd. (1999). UNC-4/UNC-37-dependent repression of motor neuron-specific genes controls synaptic choice in *Caenorhabditis elegans*. *Genes. Dev.* **13,** 2774–2786.

Xu, H., Whelan, P. J., and Wenner, P. (2005). Development of an inhibitory interneuronal circuit in the embryonic spinal cord. *J. Neurophysiol.* **93,** 2922–2933.

Yeh, E., Kawano, T., Weimer, R. M., Bessereau, J. L., and Zhen, M. (2005). Identification of genes involved in synaptogenesis using a fluorescent active zone marker in *Caenorhabditis elegans*. *J. Neurosci.* **25,** 3833–3841.

Zheng, Y., Brockie, P. J., Mellem, J. E., Madsen, D. M., and Maricq, A. V. (1999). Neuronal control of locomotion in *C. elegans* is modified by a dominant mutation in the GLR-1 ionotropic glutamate receptor. *Neuron* **24,** 347–361.

Zhou, H. M., and Walthall, W. W. (1998). UNC-55, an orphan nuclear hormone receptor, orchestrates synaptic specificity among two classes of motor neurons in *Caenorhabditis elegans*. *J. Neurosci.* **18,** 10438–10444.

MECHANOSENSATION IN *CAENORHABDITIS ELEGANS*

Robert O'Hagan* and Martin Chalfie

Department of Biological Sciences
Columbia University
New York, New York

Caenorhabditis elegans displays a variety of mechanosensory behaviors. Animals move away from both gentle and harsh touches to their bodies and gentle touches to their nose. Gentle touch also affects other behaviors, such as feeding, defecation, and egg-laying. Animals also respond to the texture of the bacterial lawns on which they feed by slowing their movement, and males initiate and proceed through mating dependent on contact with the hermaphrodite. These mechanical signals are sensed by 10% of the neurons in *C. elegans* (additional cells may act as proprioceptors). Screens for mutations affecting these behaviors have identified molecules needed for both development and function of these mechanosensory

*Current address: Massachusetts Institute of Technology, 77 Massachusetts Avenue, Cambridge Massachusetts.

INTERNATIONAL REVIEW OF
NEUROBIOLOGY, VOL. 69
DOI: 10.1016/S0074-7742(05)69006-X

0074-7742/06 $35.00

cells. These screens have yielded two types of ion channels needed for mechanosensation, the DEG/ENaC and TRP channels. Many other genes encoding cytoskeletal elements, components of the extracellular matrix, and possible accessory proteins needed for transduction have also been identified.

I. Introduction

Animals observe and process information about their environment by means of sensory neurons that transduce sensory stimuli into electrochemical signals that can be communicated to other neurons in the nervous system. Senses such as vision and olfaction have been extensively studied and their molecular mechanisms are well understood. The understanding of mechanical senses such as touch and hearing, however, has progressed more slowly. Aside from touch and hearing, animals have many other mechanical senses, such as gravitational sensing, balance, and nociception to sense external mechanical stimuli. Animals sense several internal mechanical stimuli, such as the position or location of limbs (proprioception), blood pressure (baroreception), and shear stress produced by fluid flow. Researchers are only beginning to discover the molecular mechanisms that make these senses possible.

The main reason our molecular understanding of mechanosensation has been slow to develop is that the molecules that mediate these senses are present in miniscule quantities, which precludes biochemical isolation. In the hair cells, the specialized receptors for hearing and balance in the inner ear, the molecules that transduce hearing are thought to be present only in hundreds (Howard and Hudspeth, 1988; Hudspeth, 1989), whereas <20,000 channels are thought to transduce touch in the skin of the palm and fingers of the human hand (Drummond *et al.*, 2000). A more fruitful approach has been to start with the identification of mutants defective in touch sensitivity. Genes so identified can be cloned and their products studied. This approach was first used in *C. elegans* (Chalfie and Sulston, 1981; Sulston *et al.*, 1975), but has been used subsequently in *Drosophila* (Eberl *et al.*, 1997; Kernan *et al.*, 1994) and zebrafish (Sidi *et al.*, 2003).

C. elegans devotes a significant portion of its nervous system to mechanosensation and displays a variety of behaviors in response to mechanosensory input. At least 30 of 302 neurons (about 10%) in the adult hermaphrodite are likely to be sensory mechanoreceptors and additional cells are found in males (Table I; Fig. 1). Several mechanosensory behaviors involve avoidance behaviors: gentle touch along the body, tapping the nematode culture plate, harsh touch along the body, touch to the tip of the nose, and high osmotic-strength environments all produce an avoidance response leading to the animal moving away from the stimulus. The avoidance response is likely to have several components in addition to movement. For example, gentle touch appears to modulate other sensory

TABLE I
MECHANOSENSORY NEURONS IN *C. ELEGANS*

Mechanosensory neurons	Associated behaviors
ALM (L/R), PLM (L/R), AVM, PVM	Gentle touch avoidance, resetting defecation cycle, suppressing pharyngeal pumping, suppressing foraging head movement, suppressing egg-laying
CEP (DL/DR,VL/VR), ADE (L/R), PDE (L/R)	Food-mediated slowing
ASH (L/R), FLP (L/R), OLQ (DL/DR, VL/VR)	Nose-touch avoidance, head withdrawal
ASH (L/R)	Osmotic avoidance
PVD (L/R)	Harsh touch avoidance
IL1 (DL/DR, L/R, VL/VR)	Nose-touch head withdrawal
Ventral cord motor neurons	Proprioception (?)
Putative Mechanosensory Neurons in Males	Associated behaviors
R5A, R7A, R9A (L/R; dopaminergic ray neurons)	Male-turning behavior during mating
HOA, HOB	Location of vulva by male during mating
SPC (L/R)	Insertion of spicules by male during mating
R1B, R2B, R3B, R4B, R5B, R7B, R8B, R9B (L/R; ray neurons expressing LOV-1)	Response to contact with hermaphrodite, location of vulva during mating

cells, temporarily suppresses foraging head movements and pharyngeal pumping, and resets the egg-laying and defecation cycles. Other mechanosensory modalities do not involve avoidance responses. For example, *C. elegans* senses the texture of the lawn of bacteria upon which it feeds and is also likely to sense the curvature of its own body and the tension applied by its own muscles. Additionally, males are likely to use mechanosensory cues in mating with hermaphrodites. Hence, *C. elegans* displays at least 11 different mechanosensory behavioral responses, and probably other, more subtle responses are yet to be discovered. The diversity of mechanical signals and behavioral consequences means that several different screens can be used to identify mutants with mechanosensory defects.

As in other organisms (e.g., Jan and Jan, 1993), *C. elegans* has several morphologically distinct mechanosensory cells to sense these mechanical signals. These cells can be divided into three classes according to the specialization of their cytoskeleton. The first and largest class includes neurons with ciliated sensory endings. Many of these endings abut but do not go through the cuticle of the animal. This feature has been taken as an indication that the cells are probably

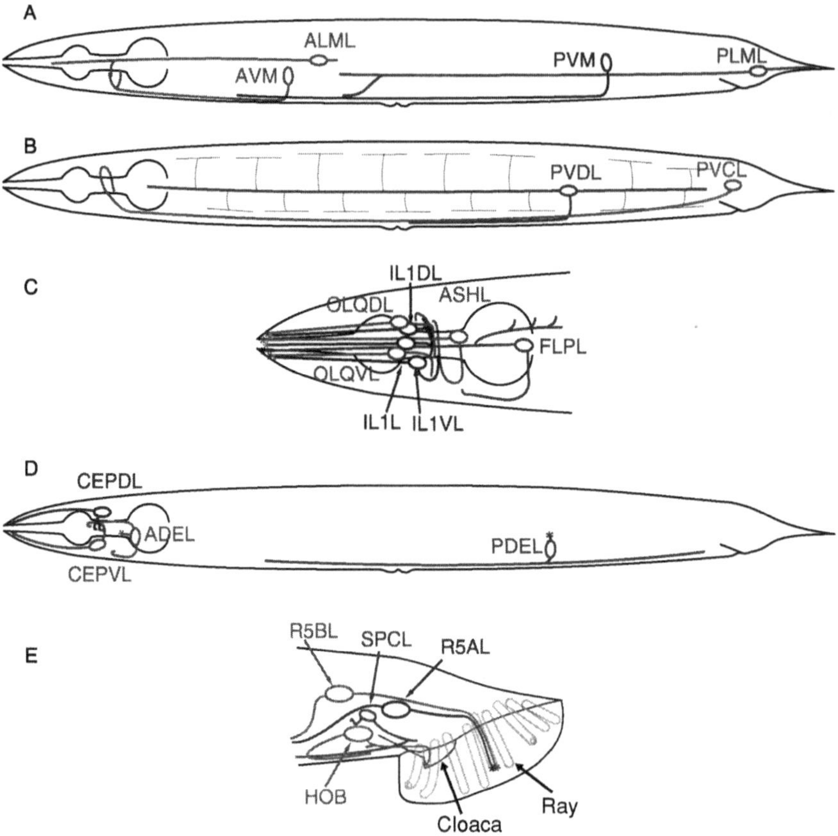

FIG. 1. Diagrams of mechanosensory neurons in *C. elegans*. A. The touch receptor neurons that sense gentle touch. When a symmetrical pair of cells exists here and in subsequent panels (e.g., ALML and ALMR) only the cell on the left side (ALML) is drawn. B. Neurons needed for harsh touch. C. Neurons that sense nose touch and osmolarity (ASH). Here and in subsequent panels, asterisks mark the positions of sensory cilia. D. Dopaminergic sensory neurons that detect texture. E. Putative mechanosensory cells of the male tail. Panels A to C were redrawn from White *et al.* (1986), panel D was redrawn from White *et al.* (1986) as modified by Tsalik *et al.* (2003), and panel E was redrawn from Emmons and Lipton (2003) and Barr *et al.* (2001). (See Color Insert.)

mechanosensory, as in Ward *et al.* (1975) for example. This characterization, however, is not a reliable rule, since the ciliated ASH neurons, which sense touch to the nose (and other signals) have their dendrites in channels that are open to the external environment (Kaplan and Horvitz, 1993; White *et al.*, 1986). The second class of cells, the touch receptor neurons that sense gentle touch (Chalfie and Sulston, 1981), do not have ciliated endings but have unusual, large-diameter microtubules in their sensory processes. The third class of cells has neurites without specialized cytoskeletons. The PVD neurons, which sense harsh touch (Way and Chalfie, 1989), are examples of this type of cell. This lack of cellular specialization makes the morphological identification of the cells as mechanosensors difficult and opens the possibility that other cells may be responding to mechanical signals. For example, R.L. Russell and L. Byerly (personal communication cited by White *et al.*, 1986) proposed that the seemingly undifferentiated regions of ventral cord motor neurons respond to body stretch. The fact that the mechanosensory sensory cells differ in their cytoskeletal specializations suggests that each may transduce mechanical stimuli by distinct molecular mechanisms.

In the following, we discuss the different types of mechanosensory behaviors by describing the sensory cells involved, their connections with the rest of the *C. elegans* nervous system, and the genes and their products that have been shown to be important for sensory function. As expected from the morphological differences, the view that is emerging is that several molecular mechanisms do underlie the sensing of mechanical signals.

II. Gentle Touch

A. Behavioral Response and Sensory Cells

When *C. elegans* is touched near the head or tail with an eyebrow hair attached to a toothpick, it reverses direction to move away from the stimulus (Chalfie and Sulston, 1981). This stimulus is sensed by six neurons that share common ultrastructural features and express many of the same genes (Fig. 1A) (Chalfie and Sulston, 1981; Chalfie and Thomson, 1979, 1982). This class of non-ciliated, but specialized neurons includes the two ALM neurons, the two PLM neurons, and the AVM and PVM neurons. The ALM and PLM cells are present at hatching, but the ventral cells (AVM and PVM) arise postembryonically (Sulston and Horvitz, 1977). All six cells have anteriorly directed processes and, except for PVM, an anterior branch. The PLM cells have a prominent posteriorly direct process. The processes of the touch neurons are embedded in the hypodermis, just beneath the cuticle, in a position that appears ideal for sensing external touch and vibrations.

These neurons were originally referred to as microtubule cells because these processes are filled with bundles of distinctive 15-protofilament microtubules (Chalfie and Thomson, 1979, 1982). They are now known collectively as the touch receptor neurons. These cells respond to a gentle touch stimulus; animals lacking the cells do not respond to the stroking of the eyebrow hair, but do respond to the harsher stimulus (as administered by prodding the animal with the platinum wire pick used to move animals [Chalfie and Sulston, 1981]). The loss of the response to gentle touch gives the Mec or *mec*hanosensory abnormal phenotype. The ALM and AVM cells respond to touch in the front half of the animals; the PLM cells respond to touch in the posterior half of the animal. Stimulating PVM alone does not produce a touch response, but its pattern of synapses (Chalfie *et al.*, 1985; see the following text) suggests that it does contribute to the touch response. Since the receptive field of a touch neuron corresponds to the entire length of the touch neuron process (Chalfie and Sulston, 1981), the entire process serves as a sensory dendrite. The process also serves as an axon conveying signals to the branches where synapses are made (White *et al.*, 1986). The processes of AVM and PVM, which are in the ventral nerve cord, also form synapses; the processes of the other touch neurons do not.

C. elegans also responds to a tap to the culture plate upon which it is grown and the same six touch receptor neurons that sense gentle touch to the body also sense plate-tap (Chalfie and Sulston, 1981). However, since the mechanical stimulus delivered in this way is non-directional, the worm cannot move away from it. Adults typically respond to a tap by reversing direction and moving backwards (Chiba and Rankin, 1990). Since the touch receptor neurons and possibly other mechanoreceptors are likely to sense this non-localized stimulus simultaneously, this type of stimulation is most useful for testing sensory integration by the larger neuronal circuit. Studies using plate-tap have often focused on modulation of the response frequency due to a form of non-associative learning termed habituation, which will be discussed later in this chapter.

The features shared by the distinctive cytoskeletal specialization and prominent extracellular matrix (ECM) of touch receptor neurons appear to be important for touch sensitivity (Fig. 2A). The large-diameter, 15-protofilament microtubules, which are linked by numerous cross bridges, fill the processes of all six touch neurons (Chalfie and Thomson, 1979, 1982). Individual microtubules are only 10 to 20 μm long, but overlap and bundle, filling the length of the process, which is 400 to 500 μm long (Chalfie and Thomson, 1979). The distal ends of the microtubules are all apposed to the plasma membrane and have associated material visible in electron micrographs that may attach these distal ends to the plasma membrane (Chalfie and Thomson, 1979, 1982). These microtubules, which are only found in the touch receptor neurons, are required for the response to touch (Chalfie and Sulston, 1981; Fukushige *et al.*, 1999; Savage *et al.*, 1994).

An electron-dense ECM called the mantle (Chalfie and Sulston, 1981) envelopes the touch receptor neuron neurites and is necessary for responses to touch. The mantle is greatest between the cell and the hypodermis underlying the body wall and periodically enlarges. The mantle is also needed to attach the touch receptor processes to the body wall. Touch receptor neurons arise adjacent to the hypodermis and are attached to it, and are later engulfed by the hypodermis during larval development. Components of the ECM are secreted by the touch neurons themselves, as well as by surrounding cells.

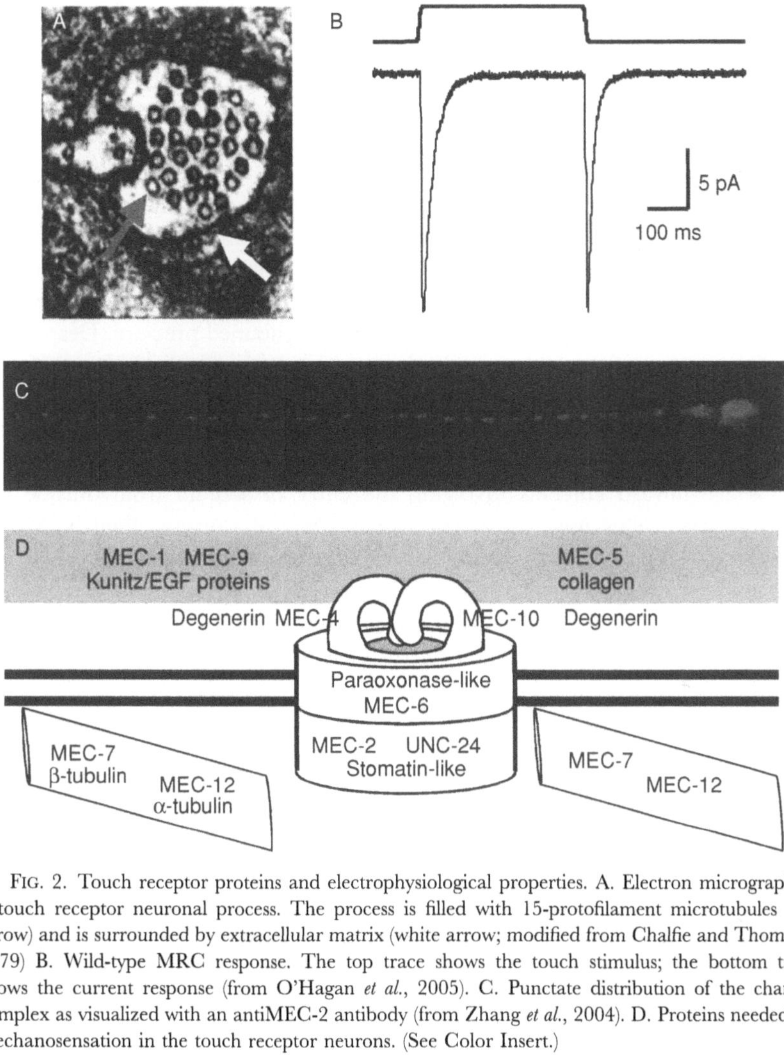

FIG. 2. Touch receptor proteins and electrophysiological properties. A. Electron micrograph of a touch receptor neuronal process. The process is filled with 15-protofilament microtubules (red arrow) and is surrounded by extracellular matrix (white arrow; modified from Chalfie and Thomson, 1979) B. Wild-type MRC response. The top trace shows the touch stimulus; the bottom trace shows the current response (from O'Hagan *et al.*, 2005). C. Punctate distribution of the channel complex as visualized with an antiMEC-2 antibody (from Zhang *et al.*, 2004). D. Proteins needed for mechanosensation in the touch receptor neurons. (See Color Insert.)

B. The Mechanoreceptor Current

Recently, electrophysiological techniques developed for recording from *C. elegans* neurons *in vivo* (Goodman *et al.*, 1998; Lockery and Goodman, 1998) have been applied to study the responses to touch in the PLM touch neurons, allowing a direct look at the response to gentle touch (O'Hagan *et al.*, 2005). The mechanoreceptor currents (or MRCs) evoked in PLM display four important characteristics (Fig. 2B). First, the MRCs activate extremely rapidly. The minimum latency of MRCs in PLM neurons is likely to be <600 μs, which is one to two orders of magnitude more rapid than known second messenger cascades (20 to 100 ms; described in Hardie and Minke, 1992). Receptor potentials observed in vertebrate hair cells and *Drosophila* mechanoreceptors are also extremely rapid (on the order of 10 μs [Corey and Hudspeth, 1979] and 200 μs [Walker *et al.*, 2000], respectively). Hence, MRCs observed in PLM recordings are likely to represent transduction by direct activation of ion channels. Second, PLM produces MRCs at the onset and offset of mechanical stimuli that rapidly adapt back to zero while the stimulus is maintained at a constant pressure; in other words, PLM responds to changes in pressure, but not maintained pressure. Despite the fact that worms habituate to repeated mechanical stimuli, displaying a decrementing magnitude and frequency of behavioral responses (Rankin *et al.*, 1990), MRCs do not appear to decrement with repeated stimuli. Thus, behavioral habituation is likely to require processes downstream of transduction. Third, MRCs are inward currents involving the entry of sodium from outside the PLM neurons. This inward current depolarizes the neurons. Fourth, MRCs are reversibly blocked by amiloride. Direct electrical recordings of the ALM cells have not been made, but calcium imaging experiments of these cells (Suzuki *et al.*, 2003) are consistent with the PLM results and suggest that responses in ALM are very similar to those observed in PLM. These experiments also indicate that calcium influx, mediated in part by the L-type voltage-gated calcium channel EGL-19, is a subsequent event in mechanosensory signaling in the touch receptor neurons.

C. Mechanosensory Reflex Circuits

The serial section reconstruction of the *C. elegans* nervous system (White *et al.*, 1986) allows an understanding of the circuitry underlying mechanosensory behaviors at the single-cell level. As first demonstrated by Chalfie *et al.* (1985), the circuitry and laser ablation can be used to deduce the functions of various cells in the nervous system. Since several mechanosensory cells modulate a common circuit for locomotion, this common circuit will be described first, followed by a description of how the touch receptor neurons interact with it.

Locomotion occurs by contractions of dorsal and ventral bodywall muscles. As a result, *C. elegans* normally lies on either its right or left side, and moves by dorsoventral undulation. By coordination of excitatory and inhibitory synaptic inputs to the muscles, contractions of the dorsal muscles are out of phase with those of the ventral muscles. The resulting alternation of these contractions produces smooth sinusoidal dorsoventral bends that propagate down the length of the worm and propel it forward or backward.

The A motor neurons (12 VA and 9 DA cells) control backward locomotion, the B motor neurons (11 VB and 7 DB cells) control forward motion, and the D motor neurons (13 VD and 6 DD cells) are required for both forward and backward movement (Chalfie *et al.*, 1985). Serial EM reconstructions indicate that the D motor neurons receive input from motor neurons of the opposite side (but no input from interneurons) and provide inhibition to the opposite side such that both dorsal and ventral muscles do not contract at the same time (White *et al.*, 1986). This role for the D motor neurons was confirmed by laser ablation studies; ablated animals shrink when touched or prodded instead of moving in a sinusoidal fashion (McIntire *et al.*, 1993).

The A and B motor neurons receive synaptic input from four pairs of interneurons—AVA, AVB, PVC, and AVD—whose processes extend the entire length of the ventral cord (White *et al.*, 1986). (The A motor neurons also receive input from a fifth pair of interneurons—AVE—whose processes are found in the anterior half of the ventral cord.) The AVB and PVC interneurons synapse electrically and chemically, respectively, onto the B motor neurons that drive forward motion (Chalfie *et al.*, 1985), and the AVA and AVD command interneurons provide input to the A motor neurons that drive backward movement. These cells are not equivalent, since ablation of PVC or AVD interneurons prevents touch-mediated responses but does not render animals uncoordinated. Ablation of the AVA or AVB interneurons produces uncoordination (the Unc phenotype). The circuitry is even more complex, since multiple electrical and chemical synapses occur between the forward and backward command interneurons. As a result, ablation suggests that the AVB and PVC also play a role in regulation of backward locomotion (Hart *et al.*, 1995).

The touch receptor neurons synapse onto all four interneurons to elicit forward and backward movement (Chalfie *et al.*, 1985). The anterior cells, ALM and AVM, form gap junctions to the AVD interneurons and the PLM neurons form gap junctions with the PVC interneurons. Ablation of AVD and PVC in young larvae results in loss of touch sensitivity for the anterior and posterior touch response, respectively (Chalfie *et al.*, 1985), suggesting that these gap junction connections are needed for the touch response. The anterior and posterior touch circuits are interconnected, since each set of touch receptor neurons form chemical synapses that are probably glutamatergic (Lee *et al.*, 1999) onto the other circuit's interneurons, i.e., ALM and AVM form chemical synapses onto PVC and PLM forms chemical synapses onto AVD. This finding suggests that the touch neurons activate one circuit while inhibiting the other (Chalfie *et al.*, 1985). This inhibition is supported by examination of the response of laser-ablated animals to a plate tap (Wicks and Rankin, 1995). The finding that plate tap results in backward movement suggest that signals from the anterior touch receptor neurons override those of posterior touch receptor neurons. This relationship between the anterior and posterior touch signals probably results from the development of the AVM neuron and its addition to the touch circuit in the last larval stage, because larval animals show fewer reversals when their plate is tapped (Chiba and Rankin, 1990; Rankin *et al.*, 2000). Chemical synapses are also made from AVM onto the AVB interneurons and onto AVA by the PLM neurons (Chalfie *et al.*, 1985). The importance of the AVM-AVB synapses is seen in animals lacking the AVD interneurons. These animals are touch insensitive at the head as young larvae, but not as older larvae (when the AVM connections are made). Additional removal of AVM results in anterior touch insensitivity at all ages.

The initial studies of the touch circuit (Chalfie *et al.*, 1985) were concerned with the backward and forward movement of the animals. In addition, anterior touch also causes the animals to reduce the right and left head oscillations that accompany foraging (Alkerna *et al.*, 2005). The ALM and AVM touch receptor neurons appear to suppress foraging head oscillations by stimulating the RIM motor neurons. These touch neurons are indirectly coupled to RIM neurons via gap junctions between the touch neurons and the AVD and AVA neurons, and from the AVD and AVA to the RIM neurons. Activation of the sensory neurons by anterior touch activates RIM neurons and causes them to secrete tyramine, which may directly inhibit contraction of the head muscles needed for head oscillations (Alkema *et al.*, 2005). Ablation of the RIM neurons (or AVA and AVD) abolishes the touch-mediated suppression of the head oscillations. Mutation of *tdc-1*, a tyrosine decarboxylase required to convert tyrosine to tyramine, also eliminates the suppression of head oscillation and causes awkward backward movement in response to touch (Alkema *et al.*, 2005).

The touch neurons synapse onto several other cells. The ALM neurons and AVM form chemical synapses onto the BDU cells and this association has been suggested as a means of guiding the outgrowth of the AVM neurons through an already established neuropil (Walthall and Chalfie, 1988). The touch neurons also synapse onto the CEP, ADE, and PDE neurons, the dopaminergic neurons that sense the texture of bacteria (see the following text). These synapses may be used for signals from the touch receptor neurons to override the texture signal and allow the animal to move away from a touch stimulus. The interaction of these two sets of mechanosensory neurons is even more complex, since humoral dopamine, presumably from the CEP, ADE, and PDE neurons, inhibits habituation of the tap response (Sanyal *et al.*, 2004). This action may be on the touch receptor neurons, which have dopamine receptors, but no direct synaptic input from these cells. The PLMR neurons chemically synapse onto the HSN neurons, which are needed for egg-laying. This connection may allow touch to inhibit egg-laying (Chalfie *et al.*, 1985). These are not the only behavioral effects of gentle touch. Touching animals inhibits pharyngeal pumping (Chalfie *et al.*, 1985) and resets the defecation rhythm (Liu and Thomas, 1994). The former behavior may involve the PVR neurons, which share gap junctions with the anterior and posterior touch receptor neurons and which synapse onto the neurons of the pharynx. The synapses needed for modulation of defecation are not known.

The picture that emerges from the analysis of the circuitry involved in the response to gentle touch is that the response is multifaceted, involving the modulation of other sensory modalities, the activation and inhibition of several different motor circuits, and the regulation of several other behavioral outputs.

D. Genes Needed for Differentiation

Saturating mutageneses have identified several genes that can mutate to cause the Mec phenotype (Chalfie and Au, 1989; Chalfie *et al.*, 1981). Even more genes have been identified through DNA microarray analysis of RNAs from the touch neurons (Zhang *et al.*, 2002). Some of these genes are needed for the development of the touch receptor neurons; the remaining genes are needed for the function of these cells.

Differentiation of the touch receptor neurons requires the UNC-86 and MEC-3 transcription factors (Chalfie *et al.*, 1981; Duggan *et al.*, 1998; Way and Chalfie, 1988, 1989; Xue *et al.*, 1992, 1993). UNC-86 is a POU-type homeodomain transcription factor that is expressed in 57 neurons (Finney and Ruvkun, 1990; Finney *et al.*, 1988). UNC-86 is needed for proper differentiation of the neuroblast precursors of the six touch receptor neurons (Chalfie *et al.*, 1981) and is expressed throughout larval growth in these cells (Finney and Ruvkun, 1990). UNC-86 activates the expression of *mec-3,* which encodes a LIM-type homeodomain transcription factor that is expressed in the touch neurons and is needed to specify their fate (Way and Chalfie, 1988). When *mec-3* activity is absent, the lineages that give rise to the touch receptor neurons are normal, but touch neuron markers are not made (Chalfie and Sulston, 1981). UNC-86 and MEC-3 interact and form a heteromultimer that binds to the promoters of *mec-3* and the genes needed for touch neuron function (Duggan *et al.*, 1998; Xue *et al.*, 1992, 1993). The heteromultimer maintains *mec-3* expression and activates the downstream function genes.

In wild-type animals, *unc-86* and *mec-3* are co-expressed in the touch neurons and two other pairs of cells—the PVD neurons and the FLP neurons (Way and Chalfie, 1989). Interestingly, these other cells are also mechanoreceptors, but they do not express the same genes as the touch receptor neurons. Both positively and negatively acting transcription factors restrict the expression of touch neuron characteristics to the six cells seen in wild-type animals (Mitani *et al.*, 1993; Toker *et al.*, 2003; Wu *et al.*, 2001).

The differentiation of the touch receptor neurons is not only controlled at the level of transcription. The touch sensitivity gene *mec-8* encodes a protein that is needed for RNA processing (Lundquist *et al.*, 1996; Yochem *et al.*, 2004). One target of *mec-8* processing is the touch function gene *mec-2* (A. Calixto, M. Huang, and M. Chalfie, 2005, unpublished data). In wild-type animals, *mec-2* produces two transcripts. The longer one, caused by the proper splicing at intron 9, is needed for *mec-2* function. This longer transcript is not made in *mec-8* mutants. Additional transcripts are probably also processed by MEC-8, since genomic *mec-2* DNA lacking intron 9 rescues a *mec-2* null mutant but not a *mec-8* null mutant.

E. Genes Needed for Function

Genetic, molecular, and electrophysiological studies have identified and characterized several genes needed for the function of the touch receptor neurons. Instead of describing all these genes, we will discuss those with a direct role in transduction or in the production of the distinct morphological features of the touch receptor neurons (Fig. 2C). In summary, transduction is mediated by a channel complex consisting of at least five proteins: the DEG/ENaC channel proteins MEC-4 and MEC-10, the stomatin-like proteins MEC-2 and UNC-24, and the paraoxonase-like protein MEC-6. This complex is localized to discrete areas in the touch neuron processes by interactions with the ECM proteins, MEC-1, MEC-5, and MEC-9. These interactions may also be needed for transduction. The 15-protofilament microtubules in the touch receptor neurons are formed from the MEC-7 and MEC-12 tubulins, but the role of the microtubules in transduction is unclear.

1. *Degenerins*

MEC-4 and MEC-10 encode similar proteins that are called degenerins because gain-of-function mutations in them can produce neuronal degeneration (Chalfie and Wolinsky, 1990; Driscoll and Chalfie, 1991; Huang and Chalfie, 1994). These proteins are members of the DEG/ENaC (*deg*enerin/*e*pithelial *Na*$^+$ *c*hannel) superfamily of ion channel proteins. In addition to a putative role in mechanosensation, other members of this superfamily have diverse roles in sodium reabsorption in the colon and kidney, clearing of liquid from the lungs, and possibly learning and memory (Kellenberger and Schild, 2002; Krishtal, 2003). Members of this superfamily have a common structure and their channels are blocked by the diuretic amiloride (Alvarez de la Rosa *et al.*, 2000; Hamill and McBride, 1996). All Deg/ENaC proteins have intracellular N- and C-termini and two transmembrane (TM) domains with a large extracellular domain containing two or three cysteine-rich domains (CRDs) between them (Kellenberger and Schild, 2002). The last CRD is the most highly conserved of the extracellular cysteine-rich domains. Because this domain has some similarity to neurotoxin peptide sequences, Tavernarakis and Driscoll (2000) have speculated that it may regulate channel gating by blocking the pore. The second transmembrane domain (TMII) is likely to form the pore of DEG/ENaC channels (Kellenberger and Schild, 2002). TMII is amphipathic, and conserved hydrophilic amino acids may line the pore lumen, since substitution of most of these residues in MEC-4 produce non-functional channels (Hong and Driscoll, 1994). However, replacing the entire TMII of MEC-4 with TMII from DEG-1 (another *C. elegans* degenerin; Chalfie and Wolinsky, 1990) or even rat α-ENaC produces a functional channel that can complement *mec-4* null alleles *in vivo* (Hong and Driscoll, 1994).

The degeneration that gives the degenerins their name can be produced by substitution of large side-chain amino acids for glycine 713 of MEC-4 (or the corresponding residue in homologs) at the beginning of the pore forming region at TMII (Chalfie and Wolinsky, 1990; Driscoll and Chalfie, 1991) or by an A404T substitution in the extracellular loop of MEC-4 can also hyperactivate channels to cause degeneration (García-Añoveros *et al.*, 1995) (these mutations are designated as *d* mutations, e.g., *mec-4d*). This latter amino acid is in a 22-amino acid sequence immediately preceding CRDII that is present only in *C. elegans* degenerins and not vertebrate superfamily members (García-Añoveros *et al.*, 1995). A small deletion in this region can also causes degeneration. Because these changes produce hyperactive DEG/ENaC channels (Kellenberger and Schild, 2002), the affected regions are likely to regulate channel gating. The hyperactivation caused by the G713 mutation is likely due to an increased open probability of the channel through steric hindrance (Driscoll and Chalfie, 1991; Hong and Driscoll, 1994; Snyder *et al.*, 2000); the hyperactivation at A404 is likely to be due to the removal of a regulatory domain (García-Añoveros *et al.*, 1995).

2. *Mechanosensory Channel Complex Proteins*

MEC-4 and MEC-10 are thought to form the pore of the mechanotransduction channel complex of the touch receptor neurons. Although genetic interactions (Gu *et al.*, 1996; Huang and Chalfie, 1994), electrophysiology and co-immunoprecipitation of proteins expressed in heterologous cells (Chelur *et al.*, 2002; Goodman *et al.*, 2002), and calcium imaging of touch neurons *in vivo* (Suzuki *et al.*, 2003) strongly suggested that MEC-4/MEC-10 channel complexes transduce touch, direct confirmation of this role has come from whole-cell voltage-clamp recordings of mechanoreceptor currents (MRCs) from PLM touch receptors *in vivo* (O'Hagan *et al.*, 2005). MRCs are absent in worms with the *mec-4(u253)* null allele. More importantly, missense mutations affecting corresponding residues in MEC-4 and MEC-10 shift the polarity of MRCs from inward to outward current responses to mechanical force, an indication that ion selectivity has been changed. This change in the property of the MRC by mutation of MEC-4 and MEC-10 (as opposed to the elimination of the current), and the rapidity of the response constitute the strongest evidence for the MEC-4/MEC-10 channel as the site of transduction. These results also indicate that MEC-4 and MEC-10 together form the pore of a heteromeric transduction channel. MEC-4 is not uniformly distributed in the cell membrane. Using a MEC-4:YFP reporter, Chelur *et al.* (2002) found that the protein is localized to discrete puncta along the touch receptor neuron process. As described in the following text, other proteins are also found in these puncta (Fig. 2D).

MEC-4 and MEC-10 are not the only components of the channel complex. Two other MEC proteins, the stomatin-like protein MEC-2 and the paraoxonase-like protein MEC-6, are also part of the complex. These proteins are required for touch sensitivity, co-immunoprecipitate with each other and with the degenerins, co-localize as puncta with MEC-4 *in vivo*, are essential for the MRCs *in vivo*, and act synergistically to increase MEC-4d currents in frog oocytes (Chalfie and Sulston, 1981; Chelur *et al.*, 2002; Goodman *et al.*, 2002; Huang *et al.*, 1995; Zhang *et al.*, 2004).

MEC-2 is a 481-amino acid integral membrane protein with unique N- and C-termini surrounding a central domain that is 65% identical to human stomatin (Huang *et al.*, 1995; Zhang *et al.*, 2004). Stomatin, a component of the red blood cell membranes, is missing in hereditary stomatocytosis, a disease with altered blood cell permeability (Stewart, 1993). Both the N- and C-termini of MEC-2 are cytoplasmic; a hydrophobic region at the beginning of the stomatin domain anchors MEC-2 to the inner leaflet of the plasma membrane (Huang *et al.* 1995). The MEC-2 N-terminus is needed for transport of the molecule in the touch neuron process (Huang *et al.*, 1995), and its C-terminus is needed for association with other MEC-2 molecules (Zhang *et al.*, 2004). MEC-2 likely associates with the MEC-4 channel complex via its stomatin-like region (Goodman *et al.*, 2002; Zhang *et al.*, 2004). This association is needed for the punctate localization of MEC-2; the absence of MEC-2 does not affect the localization of MEC-4. All three domains, however, are needed for touch sensitivity.

A second stomatin-like protein, the product of the *unc-24* gene, also appears to be part of the channel complex. In addition to a stomatin domain, UNC-24 has a non-specific lipid transfer domain (Barnes *et al.*, 1996; Sedensky *et al.*, 2001); the function of this domain is not known, but it is intriguing to speculate that it may help maintain proper lipid environment for the channel complex. UNC-24, is expressed in touch receptor neurons and co-localizes with MEC-2 puncta (Zhang *et al.*, 2004). Although *unc-24* mutants are not Mec, mutation of *unc-24* enhances the Mec phenotype caused by *mec-4* and *mec-6* temperature-sensitive alleles (Zhang *et al.*, 2004). Additionally, an HA-tagged version of UNC-24 co-immunoprecipitates with MEC-4:MYC, and co-expression of UNC-24 with MEC-4d and MEC-2 in oocytes reduces the amplitude of amiloride-sensitive current in a manner suggesting it competes with MEC-2 for participation in channel complexes (Zhang *et al.*, 2004).

MEC-6 is a 377-amino acid protein that is expressed in muscle cells, neurons (including the six touch receptor neurons), and other tissues (Chelur *et al.*, 2002). This protein is similar to human paraoxonases, proteins that have unknown functions. MEC-6 has a short cytoplasmic N-terminus, a single transmembrane domain, and a large extracellular C-terminus. Loss of *mec-6* function blocks both the degeneration caused by gain-of-function mutations in degenerin genes (Chalfie and Wolinsky, 1990; Huang and Chalfie, 1994), and the widespread expression of the gene may reflect a requirement for MEC-6 in the function of other degenerin channels. The MEC-4 puncta are absent in *mec-6* animals, suggesting that MEC-6 is needed to form the puncta or maintain them *in vivo* (a similar function is not seen when the proteins are expressed in heterologous cells; Chelur *et al.*, 2002).

Although the stoichiometry of the components of the mechanotransduction channel complex is unknown, genetic interactions suggest that several proteins are present in multiple copies. Certain alleles of both *mec-4* and *mec-10* can act as recessive gain-of-function mutations that block the degeneration caused by dominant mutations in *mec-4* (Driscoll and Chalfie, 1991; Hong and Driscoll, 1994; Hong *et al.*, 2000) or *mec-10* (Huang and Chalfie, 1994), indicating that the channel is composed of multiple MEC-4 and MEC-10 molecules. Allelic interactions among the *mec-2* alleles also suggest that multiple MEC-2 molecules interact (Chalfie and Sulston, 1981; Huang and Chalfie, 1994). In addition, YFP fluorescence of tagged MEC-4 can be seen as puncta along the touch neuron process (Chelur *et al.*, 2002), and estimates from electrophysiological data suggest that each punctum contains a single functional channel (O'Hagan *et al.*, 2005). These results suggest that multiple protein subunits are present in each punctum or that the puncta contain additional nonfunctional subunits.

3. *ECM Proteins*

Mutation of three genes encoding ECM protein—*mec-1*, *mec-5*, and *mec-9*—results in touch insensitivity. *mec-1* and *mec-9* encode proteins containing multiple EGF and Kunitz domains (Du *et al.*, 1996; Emtage *et al.*, 2004); and *mec-5* encodes a collagen that has no similarity to other collagens outside of its Gly-X-Y repeats (Du *et al.*, 1996).

In wild-type worms, touch neuron processes lie adjacent to the bodywall muscle at hatching, but move away from muscle as they are engulfed by the hypodermis as the animals mature. The processes also form hemidesmosome-like attachments to the hypodermis. Mutations in *mec-1* can cause defects in the separation of the touch neuron process and the muscle, the engulfment by hypodermal cells, and the attachment to the bodywall (Chalfie and Sulston, 1981; Emtage *et al.*, 2004). Moreover, the mantle is absent in animals with null mutations in *mec-1*. Although attachment is lost in *mec-1* animals, attachment per se is not needed for touch sensitivity. Mutations causing a C-terminal truncation of MEC-1 cause animals to be touch insensitive, but the cells are attached normally and mutations of the hemicentin gene *him-4* cause defects in attachment and engulfment of touch neuron processes, but do not produce touch-insensitivity (Emtage *et al.*, 2004; Vogel and Hedgecock, 2001).

GFP-tagged versions of MEC-1 and MEC-5 co-localize with MEC-4:YFP puncta *in vivo* (Emtage *et al.*, 2004). The formation of these ECM puncta requires *mec-1, mec-5,* and *mec-9* These ECM puncta appear to organize the channel puncta since mutations in the ECM genes prevent MEC-4 and MEC-2 puncta from forming; channel complex defects do not alter the ECM puncta (Emtage *et al.*, 2004; Zhang, 2004). The puncta align with the annuli, circumferential ring structures in the cuticle, that are produced by the hypodermis. This correlation suggests that the hypodermis regulates the distribution of puncta on the processes of the touch receptor neurons. Because the C-terminal *mec-1* mutations correctly position the MEC-5 puncta yet produce touch insensitivity, MEC-1 and perhaps other ECM proteins may serve other functions, perhaps interacting with the mature channel complex.

4. *Microtubule Proteins*

The 15-protofilament microtubule bundles that are present uniquely in touch receptor neurons are formed using the MEC-12 α-tubulin and MEC-7 β-tubulin (Chalfie and Thomson, 1982; Fukushige *et al.*, 1999; Savage *et al.*, 1989). Although null *mec-7* mutations cause complete touch insensitivity, they reduce but do not abolish MRCs (O'Hagan *et al.*, 2005), suggesting that the 15-protofilament microtubules are not essential for transduction. In fact, the role of the microtubules is unclear. Part of their function could be for transport of channel components as some defects in the distribution of these proteins is seen in *mec-7* and *mec-12* mutants (Emtage *et al.*, 2004; Huang *et al.*, 1995; Zhang, 2004).

F. Models of Mechanotransduction

Because *mec-7* and *mec-12* are touch insensitive (Chalfie and Sulston, 1981), the distal microtubule ends appear to associate with the plasma membrane (Chalfie and Thomson, 1979), and mutations in *mec-7* and *mec-12* could alter the distribution of MEC-2 (Huang *et al.*, 1995), we had originally proposed a dual-tether model for mechanosensation (Gu *et al.*, 1996). In this model, the channel complex would be tethered extracellularly to the ECM and intracellularly to the microtubules. Gating occurred when the two tethering points moved relative to each other.

Newer evidence has made this model unlikely. This newer evidence includes the findings that: (1) *mec-7* and *mec-12* null mutations do not abolish MRCs (O'Hagan *et al.*, 2005); (2) *mec-7* and *mec-12* null mutations do not prevent the formation of channel puncta (Emtage *et al.*, 2004; Zhang, 2004); and (3) MRCs occur when force is applied and removed from the animal (O'Hagan *et al.*, 2005).

These findings make the direct gating of the channel complex by coupling to the microtubules improbable. Tethering to the ECM is still possible.

G. Habituation

As previously noted, the tap-response circuit is largely the same as the gentle touch response circuit. One important difference is that the tap to the side of the worm plate, which delivers the mechanical stimulus, is likely to be administered equally and simultaneously to both anterior and posterior touch receptor neurons. In this respect, tapping may result in competition between the anterior and posterior cells (Wicks and Rankin, 1995). Tap to the plate usually causes worms to move backward rather than forward (Chiba and Rankin, 1990), and both the frequency of this response and amplitude (the distance traveled backwards) can be quantified. Amplitude decreases with the number of repeated stimuli; this decrease is known as habituation, and may be a simple form of non-associative learning (Rose and Rankin, 2001). Worms habituate more quickly to taps with a shorter interstimulus interval (ISI) and more slowly with a longer ISI, however, the memory lasts longer with longer ISI. Upon later testing, worms that have been exposed to a series of taps ("trained") have smaller-amplitude responses, and worms that had been trained with long ISI retain these smaller-amplitude responses for a longer time (Rankin and Broster, 1992). Furthermore, a series of training blocks of long ISI (i.e., 60 s) stimuli can cause long-term habituation, which worms can retain for at least 24 hours (Rose and Rankin, 2001). This long-term habituation can be blocked by exposing worms to heat-shock during their recovery between training blocks, but not before the start of training or prior to testing. Heat shock between training blocks, however, does not block accumulation of habituation over several blocks of training (Beck and Rankin, 1995); hence, long- and short-term habituation are likely to be mediated by different pathways (Rose and Rankin, 2001). Chemical cues can enhance retention of habituation (Rankin, 2000); therefore, associative conditioning is also effective in *C. elegans*.

The molecules needed for habituation are likely to be downstream of transduction by the MEC-4 channel complex and are probably involved in synaptic transmission. Habituation of transduction currents in response to touch was not observed in electrophysiological experiments *in vivo* (O'Hagan *et al.*, 2005). However, because the cell is perfused with pipette solution, these experiments do not rule out the possibility that a diffusible molecule (that might be washed out by perfusion) modifies the activity of the transduction channel to cause habituation. Additional evidence supports the idea that habituation occurs at the level of synaptic transmission: mutation of *eat-4*, a vesicular glutamate transporter, caused more rapid habituation to stimuli at several ISIs, and also caused slower recovery from habituation and abolished dishabituation of trained worms by electric shock (Rankin and Wicks, 2000). EAT-4 is likely to be needed presynaptically for normal glutamate signaling, and is expressed in touch receptor neurons (Lee *et al.*, 1999). Additionally, *glr-1* mutations cause defects in long-term retention of habituation (Rose *et al.*, 2003). Levels of this non–NMDA-type glutamate receptor may decrease in trained versus untrained wild-type worms, as visualized by GLR-1:GFP fluorescence, but the presynaptic vesicle marker, synaptobrevin (SNB-1:GFP), did not show a similar decrease. However, quantification of fluorescence of the presynaptic and postsynaptic markers was performed using different procedures and in differing areas of the ventral cord, and only the area, but not brightness, of fluorescence was quantified. Further experiments may further elucidate the ways in which glutamatergic synapses are involved in habituation to tap.

Dopamine may also be involved in habituation of touch in *C. elegans*. Worms mutant in *dop-1*, a D1-like dopamine receptor, habituate faster to tap than wild-type worms (Sanyal *et al.*, 2004). DOP-1 is expressed in PHC, ALM, and PLM neurons; CEP and PDE are thought to release dopamine humorally that may affect the touch neurons. Hence, dopamine may modulate the response of touch receptor neurons to mechanical stimulus.

III. Harsh Touch

Mutant worms that fail to respond even to prodding with a wire—a harsh touch—are phenotypically termed Tab for *t*ouch *ab*normal. In wild-type worms, the ability to detect harsh stimuli to the body is masked by the ability to detect gentle stimuli. However, in the absence of touch receptor neuron function (or absence of the touch neurons themselves by laser ablation or genetic mutation) harsh touch to the body produces an aversive locomotory response.

One harsh touch response requires the two PVD neurons (Fig. 1B), whose role as mechanosensors was first predicted by Ed Hedgecock (personal communication cited in Way and Chalfie, 1989). Animals that have the PVD neurons respond to harsh touch in the middle of the body, whereas *mec-3* mutants, where the cells do not differentiate correctly, do not (Tsalik *et al.*, 2003; Way and Chalfie, 1989). The PVD neurons extend long, undifferentiated anterior and posterior processes that are extensively branched and lack obvious synapses. This branching is missing in *mec-3* mutants (Tsalik *et al.*, 2003). The PVD cells resemble the multidendritic sensory neurons of *Drosophila*, which also have proprioceptive and nociceptive function (Adams *et al.*, 1998; Tracey *et al.*, 2003), as noted by Millet Treinin (2005, personal communication). The PVD neurons form chemical synapses with AVA and PVC (White *et al.*, 1986) that are glutamatergic (Hart *et al.*, 1995).

The *unc-86* and *mec-3* genes are expressed in the PVD neurons and are needed for their development (Chalfie *et al.*, 1981; Finney and Ruvkun, 1990; Way and Chalfie, 1989). Little is known, however, about the molecules needed for the function of the neurons. The *mec-10* gene is expressed in PVD (Huang and Chalfie, 1994), and although one might speculate that this degenerin contributes to a different touch transduction complex, *mec-10* mutations do not cause the Tab phenotype (Way and Chalfie, 1989).

mec-3 mutants, while unable to respond to gentle touch or to harsh touch through the PVD neurons, still respond to a harsh touch at the head and tail. These results suggest that several different cells can respond to harsh touch. A Tab defect in the tail is seen when the PVC interneurons (Fig. 1B) are absent. This response to harsh touch may be directly sensed by the PVC cells or by cells that synapse onto them (White *et al.*, 1986). The PVC cells express a different degenerin, the product of the *deg-1* gene and dominant gain-of-function mutations in *deg-1* cause the degeneration of the PVC interneurons and produce the Tab phenotype (Chalfie and Wolinsky, 1990). Because the loss of the gene does not produce this phenotype (the animals appear to be wild-type), the role of this degenerin in PVC function is unknown (Chalfie and Wolinsky, 1990).

IV. Osmolarity

C. elegans is repelled by regions of high osmolarity and several genes have been identified for this avoidance behavior (Culotti *et al.*, 1978). Because high osmolarity is sensed by the ASH neurons, the discussion of these cells and genes is included in the following section on nose touch.

V. Nose Touch

A. Behavioral Response and Sensory Cells

When animals are touched at the tip of the nose or bump into objects, they respond by backing away from the stimulus or moving the head (Croll, 1976). Worms that fail to respond to gentle touch to the tip of the nose have the phenotype known as Not for *no*se *t*ouch defective. This response occurs in the absence of the touch receptor neurons needed for gentle touch (Chalfie and Sulston, 1981).

Three sets of neurons sense touch to the nose of the animal: two ASH, two FLP, and four OLQ sensory neurons (Fig. 1C) (Kaplan and Horvitz, 1993).

These ciliated neurons located in the head have cell bodies in the nerve ring and extend processes to the tip of the nose (White *et al.*, 1986). These cells may mediate responses in different receptive fields for touch around the nose, since worms touched on the dorsal side of the nose differ in their response from those touched on the ventral side (J. M. Kaplan and H. R. Horvitz, personal communication cited in Hart *et al.*, 1995). Sensation of nose-touch by multiple classes of neurons may simply increase sensitivity. Removal of the ASH neurons reduces nose touch by 45%, removal of the FLP neurons reduces it by 30%, and removal of the OLQ neurons reduces it by 5% (Kaplan and Horvitz, 1993). An animal with all three classes of cells ablated still responds to nose-touch 10% of the time. This remaining response may be mediated by non-amphid neurons in the head (Kaplan and Horvitz, 1993) or anterior touch receptor neurons (Driscoll and Kaplan, 1997).

The nose-touch receptor neurons activate the locomotion circuitry through several connections. ASH is electrically coupled to AVD, and both ASH and FLP form glutamatergic synapses onto the AVA, AVB, and AVD interneurons (Baran *et al.*, 1999; Hart *et al.*, 1995, 1999; Maricq *et al.*, 1995); OLQ is synaptically connected to these neurons indirectly, via a connection to RIC (White *et al.*, 1986).

A head-withdrawal response also occurs in response to nose-touch. Laser ablation studies indicate that the OLQ neurons, along with IL1 ciliated motor neurons, may mediate the aversive head-withdrawal reflex in response to touch on the dorsal or ventral side of the nose (Hart *et al.*, 1995). The OLQ and IL1 neurons form glutamatergic synapses onto RMD motor neurons that control muscles for head movement. OLQ and IL1 seem to mediate spontaneous foraging (head movement) as well, since animals lacking these cells (via laser ablation) forage abnormally slowly (Hart *et al.*, 1995).

B. Genes Needed for Differentiation

Both the kinesin OSM-3 (Tabish *et al.*, 1995) and the dynein heavy chain protein CHE-3 (Wicks *et al.*, 2000) are required for proper formation of ciliated endings of many neurons, including the ciliated mechanosensory neurons needed for nose touch. Mutations in several other genes cause defects in ciliated neurons (e.g., Starich *et al.*, 1995); these genes are probably needed in general for formation or function of ciliated endings. Transcription factor regulating the fate of these cells are not known except for the FLP neurons. These cells express UNC-86, MEC-3, and LIN-14, which would normally promote the production of touch receptor neurons, but this action is prevented by two other transcription factors EGL-44 and EGL-46 (Finney and Ruvkun, 1990; Way and Chalfie, 1989; Wu *et al.*, 2001).

C. Genes Needed for Function

1. *TRP-Like Ion Channel Components*

Aside from the DEG/ENaC proteins, the major family of channel proteins implicated in sensory mechanotransduction is the TRP-like proteins. TRP proteins are so named because the first member identified, needed for sustained photoreceptor responses, was called *t*ransient *r*eceptor *p*otential (Hardie and Minke, 1992). TRP proteins have six transmembrane segments and form non-specific cation channels; their permeability to calcium may be important for their function (Montell, 2003). Involvement of TRP proteins in mechanosensation was first suggested by the identification of the product of the *C. elegans osm-9* gene, a gene needed for ASH function, as a member of the family (Colbert *et al.*, 1997). Subsequent experiments in *Drosophila* identified the involvement of TRP-like channel genes in the functions of mechanosensory bristles (*nompC*; Walker *et al.*, 2000), auditory receptors (*iav* and *nan*; Gong *et al.*, 2004), and nociceptors (*painless*; Tracey *et al.*, 2003). Two TRP proteins, a NompC-like protein and TRPA1, are needed for hair cell function in zebrafish (Corey *et al.*, 2004; Sidi *et al.*, 2003). A mammalian homolog of NompC has not been discovered, but TRPA1 is needed for mammalian hair cell function (Corey *et al.*, 2004; Nagata *et al.*, 2005).

The ASH neurons mediate aversive responses to several noxious stimuli; in addition to nose-touch, they also respond to high osmolarity and noxious chemicals (Bargmann *et al.*, 1990; Hart *et al.*, 1995; Hilliard *et al.*, 2005; Troemel *et al.*, 1995). The TRPV genes *osm-9* and *ocr-2* are needed for all these responses (Colbert *et al.*, 1997; Tobin *et al.*, 2002). GFP-tagged versions of OCR-2 and OSM-9 are likely to co-assemble into channels that localize to cilia (Tobin *et al.*, 2002). OSM-9::GFP is expressed in ASH, FLP, OLQ, and PVD mechanosensory neurons. The TRPV genes *osm-9* and *ocr-1* through *ocr-4* are expressed in overlapping patterns that suggest fine-tuning of responses via heteromultimerization (Tobin *et al.*, 2002).

The general requirement of the TRP-channel proteins in these cells suggests that if they do transduce mechanical signals, other proteins must give them the required specificity. In fact, the different sensory modalities detected in ASH are genetically separable. The *osm-10* gene, which is expressed in ASH, is needed for osmosensation, but not mechanosensation (Hart *et al.*, 1999). Mechanosensation can also be lost without a loss in osmosensation (see the following text).

Addition insights have been obtained recently by the use of optical calcium indicators to monitor ASH. Using the cameleon calcium sensor, Hilliard *et al.* (2005) showed that touch elevated cytosolic calcium in the presence of exogenous serotonin and wild-type *osm-9*. The other sensory modalities of ASH did not require exogenous serotonin and were not modulated by its presence. (The osmotic responses of ASH have been visualized using another genetically encoded calcium indicator, G-CaMP [Kahn-Kirby *et al.*, 2004].) The wild-type response to nose-touch in ASH is likely to have a very low conductance, and other factors regulated by serotonin may have a role in the response.

Expression of a mammalian TRPV4 protein in ASH can rescue defects in osmotic and nose-touch avoidance in *osm-9* mutants (Liedtke *et al.*, 2003). TRPV4-mediated rescue of osmosensitivity requires activity of OCR-2, ODR-3 (a Gα), and OSM-10 (a cytosolic protein) and rescue of nose-touch requires OCR-2 and ODR-3. In mammals, TRPV4 is expressed in circumventricular organs, which detect minute osmolarity changes in the brain; TRPV4 may directly transduce this type of mechanical stimulus, but the genetic requirements for rescue in *C. elegans* may suggest that other components such as G-protein–coupled receptors may mediate transduction (Liedtke *et al.*, 2003). These differing requirements also point out that the sensory functions of the ASH polymodal neurons are genetically separable (Hart *et al.*, 1999; Liedtke *et al.*, 2003; Tobin *et al.*, 2002).

2. *Other Components*

The first gene shown to be needed for the nose-touch response was the glutamate receptor gene *glr-1*, which is expressed in the postsynaptic partners of ASH (Hart *et al.*, 1995; Maricq *et al.*, 1995). Further work on decoding ASH responses in different sensory modes has shown that ASH releases glutamate synaptically in response to all three modes, but osmotic avoidance requires the activity of the NMDA-type glutamate receptor, NMR-1, and the non-NMDA receptors, GLR-1 and GLR-2, while nose-touch avoidance requires only the non-NMDA receptors (Mellem *et al.*, 2002). EAT-4 is required for responses in all three modes. A glutamate-gated chloride conductance is also present in the AVA interneurons postsynaptic to ASH, which may serve to re-establish forward movement after the noxious stimulus has ended (Mellem *et al.*, 2002). Loss of function of EGL-3 proprotein convertase can actually suppress the Not phenotype in *glr-1* mutants in an *nmr-1*–dependent manner (Mellem *et al.*, 2002) via changes that may be presynaptic, as exogenous application of glutamate to AVA produced no difference in electrophysiological responses of wild-type or *egl-3* worms. The normal role of EGL-3 may be to downregulate glutamate concentration in ASH; hence, loss of EGL-3 may increase glutamate release, and in *glr-1* mutants, this may cause increased activation of the glutamate-gated chloride conductance.

Several TRP channels are activated by a G-protein–coupled receptor and the tyrosine-kinase second-messenger pathways, and pathways such as these may influence OSM-9 (Kahn-Kirby *et al.*, 2004). Polyunsaturated fatty acid (PUFA) levels can diminish or abolish behaviors mediated by ASH and OSM-9. OSM-9–dependent calcium transients triggered by high osmolarity and visualized by G-CaMP were influenced by the presence of PUFAs; the concentration of PUFAs could be modified by mutation of genes involved in lipid metabolism or by exogenous application. Exogenous application of PUFAs increased cytosolic calcium without additional stimuli. This finding may highlight the importance of lipid to TRP function as well. Whether these fatty acids influence OSM-9 by inclusion into membrane lipid or by increasing concentrations of known lipid second messengers such as PIP_2, IP_3, or DAG is unknown.

VI. Texture-Mediated Slowing

C. elegans also detects textural differences in its substrate, such as would be experienced when crawling from an area of the culture plate agar devoid of bacteria to an area with bacteria. *C. elegans* eats bacteria, and chemotaxes towards this food source using olfactory cues; but additionally, it responds to the difference in texture by slowing its locomotion when entering a bacterial lawn. This behavior is indeed mechanosensory, since worms entering a faux bacterial lawn of sterile G-200 Sephadex beads slow similarly (Sawin *et al.*, 2000). This behavioral response serves to keep the animals in the midst of food.

The CEP, ADE, and PDE neurons are the sensory cells required for this food-induced slowing (Fig. 1D). They have ciliated sensory endings that are embedded in the cuticle for apparent mechanosensitivity (Perkins *et al.*, 1986; Ward, 1975; White *et al.*, 1986). Laser ablation confirmed that these neurons are required for this response to small particles (Sawin *et al.*, 2000). Ablation of the four CEP neurons reduced the degree of slowing on food or beads, and additional ablation of the ADE and PDE on only one side of the animal affected slowing only when the animal was lying on the operated side (Sawin *et al.*, 2000). The four CEP neurons extend processes to the tip of the nose; the ADE neurons are bilaterally paired cells in the anterior; and the PDE neurons are bilaterally paired cells in the posterior (Sulston *et al.*, 1975). These eight neurons are the only dopaminergic neurons in the hermaphrodite, and food-induced slowing also requires dopamine (Sawin *et al.*, 2000).

Upon mechanical detection of the texture of food, the sensory neurons must deliver an inhibitory signal to the motor circuit, since the worm slows or stops. The CEP cells synapse onto the AVE and the RIC interneurons which enter the motor circuit via chemical synapses to AVA (White *et al.*, 1986). The ADE neurons chemically synapse onto the CEP sensory neurons, and also the AVA interneurons to enter the motor circuit (White *et al.*, 1986). PDE neurons synapse electrically with PVC interneuron and chemically with DVA interneurons to influence the motor circuit. Synapses from other mechanosensory neurons may affect the function of CEP, ADE, and PDE. In addition to the interactions with the touch receptor neurons previously noted, these cells receive electrical or chemical synapses from other mechanosensory neurons (FLP, OLQ, and IL2) (White *et al.*, 1986).

The response to texture differs according to the feeding state of the worm (Sawin *et al.*, 2000). Although the well-fed and starved approach a bacterial lawn similarly and both worms are slow when entering the lawn, starved worms will slow their movement to a much greater extent. The slowing response of well-fed animals depends on dopamine, occurs on Sephadex beads, and requires the CEP, ADE, and PDE neurons; the slowing of starved animals depends on serotonin, but does not occur on the beads or require the dopaminergic neurons. Thus, the response to texture only appears to occur when the animals are well-fed, and starvation leads to the inhibition of CEP, ADE, and PDE neurons.

The molecules that transduce textural mechanosensory stimuli in the CEP, ADE, and PDE neurons are unknown, but the closest *C. elegans* homolog of the *Drosophila nompC* gene (*trp-4*) is expressed in CEP and ADE neurons (Walker *et al.*, 2000). An allele with a deletion and insertion in *trp-4* was recently generated by the *C. elegans* Knockout Consortium (*www.wormbase.org*). Assays for the texture response and electrophysiological analysis of mutant animals should determine whether this protein transduces mechanical stimuli.

VII. Proprioception and Sinusoidal Locomotion

The sensory behaviors and phenotypes previously listed are largely involved with the worm's ability to sense mechanical stimuli from the environment. However, worms may have a form of proprioception that senses forces arising in the worm itself as the worm undulates to achieve locomotion. The VA, VB, DA, and DB motor neurons that regulate muscular contractions needed for locomotion may sense their own degree of curvature (Tavernarakis *et al.*, 1997; White *et al.*, 1986). The long, undifferentiated regions of these neurons that do not receive synaptic input or participate in neuromuscular junctions have been proposed by R.L. Russell and L. Byerly (personal communication cited by White *et al.*, 1986) to be stretch sensitive, as have regions distal to the neuromuscular junctions (Tavernarakis *et al.*, 1997). Stretch sensation by these neurons may provide sensory feedback of the posture of the worm's body to coordinate the timing of alternating contractions and relaxations that allow the worm to move efficiently.

Proprioception may be mediated by the degenerins UNC-8 and DEL-1 (Tavernarakis *et al.*, 1997), as both are expressed in the VA and VB motor neurons, as well as the FLP sensory neurons (Tavernarakis *et al.*, 1997). This co-expression leaves open the possibility that, like other DEG/ENaCs, UNC-8 and DEL-1 may co-assemble to form a heteromeric channel. Loss-of-function and putative null alleles of *unc-8* cause an uncoordinated phenotype in which the sinusoidal body waves produced by locomotion are reduced in both amplitude and wavelength (Tavernarakis *et al.*, 1997).

The assembly of a DEG/ENaC ion channel complex in motor neurons that is similar to the MEC-4 complex in the touch receptor neurons is supported by several observations. First, a null allele of *mec-6* also reduces wavelength and amplitude of sinusoidal body movements. This effect is not mediated by activity of MEC-6 in the MEC-4 transduction complex in the touch receptor neurons, since *mec-4* null worms have normal sinusoidal motions (Tavernarakis *et al.*, 1997). Loss of *mec-6* also suppresses uncoordination induced by an *unc-8* gain-of-function allele (Shreffler *et al.*, 1995). Loss of UNC-1, another stomatin homolog, can also suppress uncoordination produced by a neomorphic allele (Rajaram *et al.*, 1999), and UNC-1 co-immunoprecipitates UNC-8 in worm extracts (Sedensky *et al.*, 2004). Hence, inclusion of MEC-6 and stomatin homologs in an UNC-8 channel complex may be another representative of a general DEG/ENaC-paraoxonase-stomatin system. However, mechanosensitivity of this channel complex has not been demonstrated. Electrophysiology *in vivo* from motor neurons may prove illuminating, although stimulation of a proprioception channel may entail technical challenges.

Additionally, bodywall muscles may sense their own degree of stretch in order to coordinate contraction for organized locomotion, however, evidence for this is minimal. Another degenerin, *unc-105*, is expressed in *C. elegans* muscle, and its loss of function causes uncoordinated movement. A dominant allele causes hypercontraction but not death (García-Añoveros *et al.*, 1998). However, efforts to mechanically induce opening of the wild-type UNC-105 channel while recording whole-cell currents *in vivo* failed to reveal any mechanosensitivity (Jospin *et al.*, 2004).

VIII. Male Mating and Location of the Vulva

The complicated sequence of movements required for male-mating behavior is likely to utilize both chemosensory and mechanosensory cues (Liu and Sternberg, 1995). Mating begins when a male encounters a hermaphrodite, usually by bumping into it. The male maintains contact using its tail while sliding along the hermaphrodite body in a search for the vulva (Emmons and Sternberg, 1997). The male identifies the vulva by cues that are probably chemosensory and mechanosensory, and inserts its spicules and then sperm. If the male has not located the vulva by the time it has reached the end of the hermaphrodite, it makes a sharp ventral flex of the tail to turn around the end to continue sliding along the hermaphrodite. This turning behavior may require cues involving touch such as mechanical detection of tapering at the ends of the body to maintain contact with its partner and proprioception to produce the sharp turn. However, end-specific chemosensory cues may also be important for the turning behavior (Emmons and Sternberg, 1997).

A. Sensory Neurons and Circuitry

Males have 87 neurons not found in the hermaphrodite and approximately one-half of these are ciliated neurons and are thought to be sensory (Fig. 1E) (Sulston *et al.*, 1980). These neurons innervate the nine pairs of tail rays, the hook, the post-cloacal sensilla, and the spicules. Each ray has two ciliated neurons. The ray neurons trigger a mating response upon contact with a hermaphrodite, and the posterior three pairs of rays are needed to execute the turn that males make near the end of the hermaphrodite. Only the neurons of the sixth ray are not exposed to the outside environment (Sulston *et al.*, 1980) and so may be mechanosensory. Some other ray neurons may be mechanosensory because, like the ADE, PDE, and CEP neurons, they are dopaminergic. Six ray neurons (R5A, R7A, and R9A neuron pairs) are dopaminergic, and males lacking the ability to produce dopamine mate less efficiently than wild-type (Hodgkin, 1983). Animals lacking the dopaminergic neurons fail to properly execute the sharp turn (Liu and Sternberg, 1995). Mechanosensory functions may be provided by other neurons, such as the HOA and HOB hook neurons, the spicules and post-cloacal sensillae, which are needed for detection of the vulval opening (Barr and Sternberg, 1999). The SPD sensory neurons and the SPC motor neurons, which have been proposed to possess proprioceptive function (Sulston *et al.*, 1980), are required for spicule insertion (Liu and Sternberg, 1995). However, whether these sensory neurons use mechanical, chemical, or both types of sensory cues for this task is unknown.

Although several sensory cells specific to the male tail may be mechanosensors, the connections made by these cells to the rest of the nervous system are unknown, because unlike the hermaphrodite nervous system, the male nervous system has not been fully reconstructed from serial electron micrographs. Some cells involved in the turning movement of the male along the hermaphrodite's body include the EF interneurons, which receive synaptic inputs from the ray neurons (Emmons and Sternberg, 1997) and the CP motor neurons (Loer and Kenyon, 1993).

B. Putative Mechanotransducers in Male Neurons

Studies of genes needed for male mating have identified two genes that may transduce mechanical signals: *lov-1* and *pkd-2* (Barr and Sternberg, 1999). *lov-1* mutants were obtained because males had a Lov (defective *l*ocation *o*f *v*ulva) phenotype. These animals can perform many of the movements needed for male-mating behavior, but still fail to locate the vulva. They also often fail to initiate mating behavior upon contact with a hermaphrodite. Tagged versions of LOV-1 and PKD-2 are expressed in male tail ray neurons and hook neurons as well as the male-specific CEM neurons in the head, and co-localize in cilia and cell bodies. Ciliary localization is necessary for function of these proteins (Barr and Sternberg, 1999; Barr *et al.*, 2001).

LOV-1 and PKD-2 are homologs of human PKD-1 and PKD-2, respectively, which have been implicated in hereditary polycystic kidney disease. These two proteins are thought to act as a receptor (PKD-1)-channel (PKD-2) complex. PKD-1 has features of G-protein–coupled receptors and PKD-2 is a TRP channel protein (Delmas *et al.*, 2002; Emmons and Somlo, 1999; Nauli and Zhou, 2004). These proteins are localized to the primary cilium of the kidney epithelium and are needed to sense fluid shear stress in cultured kidney epithelial cells (Nauli and Zhou, 2004; Nauli *et al.*, 2003).

LOV-1 lacks the intracellular coiled-coil tail domain at the C-terminus that is the site of interaction of mammalian PKD-1 with PKD-2; however, C-terminal truncations in LOV-1 still destroy function (Barr and Sternberg, 1999). Furthermore, phenotypes of *pkd-2* and *lov-1;pkd-2* double mutants are identical to single mutants, indicating that these genes act in the same pathway (Barr *et al.*, 2001). Hence, despite the absence of a known interaction domain in LOV-1, it and PKD-2 are likely to assemble into a complex that forms a calcium-permeable ion channel like their mammalian counterparts (Barr *et al.*, 2001; Emmons and Somlo, 1999; Hanaoka *et al.*, 2000; Qian *et al.*, 1997; Tsiokas *et al.*, 1997).

Two other proteins whose human counterparts are implicated in human kidney disease, NPHP-1 and NPHP-4, may connect LOV-1 and PKD-2 mechanotransduction channels to axonemes of the cilia in male sensory neurons (Jauregui and Barr, 2005). NPHP-1 is 42% similar to human nephrocystin-1 and contains coiled-coil domains, a glutamate-rich domain, a glutamine- and proline-rich domain, and an SH3 interaction domain. Little is known about NPHP-4, which lacks other motifs for which function has been described. In *C. elegans,* these proteins appear to co-localize with PKD-2 in cilia of HOB, ray neurons and neurons of the postcloacal sensillae. Double mutants of *nphp-1* and *nphp-4* (but not single mutants) also have defects in turning and vulva location.

The localization of LOV-1 and PKD-2 to the cilia is important for their function. A kinesin-3 homolog, KLP-6, is needed for this ciliary localization. Peden and Barr (2005) have proposed that this protein may both transport and anchor LOV-1/PKD-2 complexes to the cytoskeleton, perhaps even acting similar to myosin-1c in vertebrate hair cells for mechanosensory adaptation.

The LOV-1/PKD-2 TRP channel complex may not be the only possible transducer of touch in male-specific neurons. Another putative ion channel mechanotransducer, *twk-9*, is expressed in one of the pair of neurons in each ray except ray 6 (most likely the RnA neurons) (Portman and Emmons, 2004). TWK-9 is a two-pore potassium channel. This class of potassium channel can be mechanically activated in membrane patches (Bang *et al.*, 2000; Patel *et al.*, 1998). Little is known about the function of the two-pore potassium channels, and a physiological role in mechanosensation has not been demonstrated.

IX. Conclusions and Future Directions

This is an exciting time for investigation of mechanosensation using *C. elegans*. The genetic manipulation possible in *C. elegans* has long been fruitful in the study of mechanosensation and is still a powerful tool. However, powerful new techniques, such as direct electrophysiological recording from neurons, imaging of genetically-encoded calcium sensors, and detection of gene expression using microarrays, are becoming more commonplace and are already yielding excellent results. Although direct evidence that a MEC-4/MEC-10 channel complex transduces touch exists, *C. elegans* has 24 other predicted DEG/ENaC family members and 24 putative TRP channels (Goodman and Schwartz, 2003). The functions of these proteins as mechanotransducers can also now be tested by direct electrophysiological observation and in knock-out mutants. With these new tools and classical and molecular genetics, we should be able to learn the answers to long-standing questions about the mechanisms of mechanosensation in the near future. Some of the questions that need to be addressed are: (1) What do mechanotransduction channels and their gating domains look like? (2) Does transduction occur because gating domains act like trap doors that swing open and shut or does the channel act more like the iris of a camera, twisting to constrict a central pore? (3) How does force activate gating, and how do putative extracellular tethers transfer force to these channels? and (4) If tethers to specialized cytoskeletal elements are not required for gating (as suggested by the remaining transduction current in PLM cells lacking large-diameter microtubules), what is the role of the cytoskeleton in mechanosensation? Much still remains to be uncovered about mechanosensation in *C. elegans*.

Acknowledgments

Work from our laboratory is supported by N.I.H. grant GM30997 to M. C. and a Howard Hughes Predoctoral Fellowship to R. O.

References

Adams, C. M., Anderson, M. G., Motto, D. G., Price, M. P., Johnson, W. A., and Welsh, M. J. (1998). *Ripped pocket* and *pickpocket*, novel *Drosophila* DEG/ENaC subunits expressed in early development and in mechanosensory neurons. *J. Cell Biol.* **140,** 143–152.

Alkema, M. J., Hunter-Ensor, M., Ringstad, N., and Horvitz, H. R. (2005). Tyramine functions independently of octopamine in the *Caenorhabditis elegans* nervous system. *Neuron* **46,** 247–260.

Alvarez de la Rosa, D., Canessa, C. M., Fyfe, G. K., and Zhang, P. (2000). Structure and regulation of amiloride-sensitive sodium channels. *Annu. Rev. Physiol.* **62,** 573–594.

Bang, H., Kim, Y., and Kim, D. (2000). TREK-2, a new member of the mechanosensitive tandem-pore K^+ channel family. *J. Biol. Chem.* **275,** 17412–17419.

Baran, R., Aronoff, R., and Garriga, G. (1999). The *C. elegans* homeodomain gene *unc-42* regulates chemosensory and glutamate receptor expression. *Development* **126,** 2241–2251.

Bargmann, C. I., Thomas, J. H., and Horvitz, H. R. (1990). Chemosensory cell function in the behavior and development of *Caenorhabditis elegans*. *Cold Spring Harb. Symp. Quant. Biol.* **55,** 529–538.

Barnes, T. M., Jin, Y., Horvitz, H. R., Ruvkun, G., and Hekimi, S. (1996). The *Caenorhabditis elegans* behavioral gene *unc-24* encodes a novel bipartite protein similar to both erythrocyte band 7.2 (stomatin) and nonspecific lipid transfer protein. *J. Neurochem.* **67,** 46–57.

Barr, M. M., De Modena, J., Braun, D., Nguyen, C. Q., Hall, D. H., and Sternberg, P. W. (2001). The *Caenorhabditis elegans* autosomal dominant polycystic kidney disease gene homologs *lov-1* and *pkd-2* act in the same pathway. *Curr. Biol.* **11,** 1341–1346.

Barr, M. M., and Sternberg, P. W. (1999). A polycystic kidney-disease gene homologue required for male mating behaviour in *C. elegans*. *Nature* **401,** 386–389.

Beck, C. D., and Rankin, C. H. (1995). Heat shock disrupts long-term memory consolidation in *Caenorhabditis elegans*. *Learn. Mem.* **2,** 161–177.

Chalfie, M., and Au, M. (1989). Genetic control of differentiation of the *Caenorhabditis elegans* touch receptor neurons. *Science* **243,** 1027–1033.

Chalfie, M., Horvitz, H. R., and Sulston, J. E. (1981). Mutations that lead to reiterations in the cell lineages of *C. elegans*. *Cell* **24,** 59–69.

Chalfie, M., and Sulston, J. (1981). Developmental genetics of the mechanosensory neurons of *Caenorhabditis elegans*. *Dev. Biol.* **82,** 358–370.

Chalfie, M., Sulston, J. E., White, J. G., Southgate, E., Thomson, J. N., and Brenner, S. (1985). The neural circuit for touch sensitivity in *Caenorhabditis elegans*. *J. Neurosci.* **5,** 956–964.

Chalfie, M., and Thomson, J. N. (1979). Organization of neuronal microtubules in the nematode *Caenorhabditis elegans*. *J. Cel. Biol.* **82,** 278–289.

Chalfie, M., and Thomson, J. N. (1982). Structural and functional diversity in the neuronal microtubules of *Caenorhabditis elegans*. *J. Cell Biol.* **93,** 15–23.

Chalfie, M., and Wolinsky, E. (1990). The identification and suppression of inherited neurodegeneration in *Caenorhabditis elegans*. *Nature* **345,** 410–416.

Chelur, D. S., Ernstrom, G. G., Goodman, M. B., Yao, C. A., Chen, L., R, O. H., and Chalfie, M. (2002). The mechanosensory protein MEC-6 is a subunit of the *C. elegans* touch-cell degenerin channel. *Nature* **420,** 669–673.

Chiba, C. M., and Rankin, C. H. (1990). A developmental analysis of spontaneous and reflexive reversals in the nematode *Caenorhabditis elegans*. *J. Neurobiol.* **21,** 543–554.

Colbert, H. A., Smith, T. L., and Bargmann, C. I. (1997). OSM-9, a novel protein with structural similarity to channels, is required for olfaction, mechanosensation, and olfactory adaptation in *Caenorhabditis elegans*. *J. Neurosci.* **17,** 8259–8269.

Corey, D. P., García-Añoveros, J., Holt, J. R., Kwan, K. Y., Lin, S. Y., Vollrath, M. A., Amalfitano, A., Cheung, E. L., Derfler, B. H., Duggan, A., Geleoc, G. S., Gray, P. A., Hoffman, M. P., Rehm, H. L., Tamasauskas, D., and Zhang, D. S. (2004). TRPA1 is a candidate for the mechanosensitive transduction channel of vertebrate hair cells. *Nature* **432,** 723–730.

Corey, D. P., and Hudspeth, A. J. (1979). Response latency of vertebrate hair cells. *Biophys. J.* **26,** 499–506.

Croll, N. A. (1976). When *Caenorhabditis elegans* (Nematoda: Rhabditidae) bumps into a bead. *Can. J. Zool.* **54,** 566–570.

Culotti, J. G., and Russell, R. L. (1978). Osmotic avoidance defective mutants of the nematode *Caenorhabditis elegans*. *Genetics* **90,** 243–256.

Delmas, P., Nomura, H., Li, X., Lakkis, M., Luo, Y., Segal, Y., Fernandez-Fernandez, J. M., Harris, P., Frischauf, A. M., Brown, D. A., and Zhou, J. (2002). Constitutive activation of G-proteins by polycystin-1 is antagonized by polycystin-2. *J. Biol. Chem.* **277,** 11276–11283.

Driscoll, M., and Chalfie, M. (1991). The *mec-4* gene is a member of a family of *Caenorhabditis elegans* genes that can mutate to induce neuronal degeneration. *Nature* **349,** 588–593.

Driscoll, M., and Kaplan, J. (1997). Mechanotransduction. *In* "*C. elegans* II" (D. L. Riddle, T. Blumenthal, B. J. Meyer, and J. R. Priess, Eds.), pp. 645–677. Cold Spring Harbor Laboratory Press, Cold Spring Harbor, New York.

Drummond, H. A., Abboud, F. M., and Welsh, M. J. (2000). Localization of β and γ subunits of ENaC in sensory nerve endings in the rat foot pad. *Brain Res.* **884,** 1–12.

Du, H., Gu, G., William, C. M., and Chalfie, M. (1996). Extracellular proteins needed for *C. elegans* mechanosensation. *Neuron* **16,** 183–194.

Duggan, A., Ma, C., and Chalfie, M. (1998). Regulation of touch receptor differentiation by the *Caenorhabditis elegans mec-3* and *unc-86* genes. *Development* **125,** 4107–4119.

Eberl, D. F., Duyk, G. M., and Perrimon, N. (1997). A genetic screen for mutations that disrupt an auditory response in *Drosophila melanogaster*. *Proc. Natl. Acad. Sci. USA* **94,** 14837–14842.

Emmons, S. W., and Somlo, S. (1999). Signal transduction. Mating, channels and kidney cysts. *Nature* **401,** 339–340.

Emmons, S. W., and Sternberg, P. W. (1997). Male development and mating behavior. *In* "*C. elegans* II" (D. L. Riddle, T. Blumenthal, B. J. Meyer, and J. R. Priess, Eds.), pp. 295–334. Cold Spring Harbor Laboratory Press, Cold Spring Harbor, New York.

Emmons, S. W., and Lipton, J. (2003). Genetic basis of male sexual behavior. *J. Neurobiol.* **54,** 93–110.

Emtage, L., Gu, G., Hartwieg, E., and Chalfie, M. (2004). Extracellular proteins organize the mechanosensory channel complex in *C. elegans* touch receptor neurons. *Neuron* **44,** 795–807.

Finney, M., and Ruvkun, G. (1990). The *unc-86* gene product couples cell lineage and cell identity in *C. elegans*. *Cell* **63,** 895–905.

Finney, M., Ruvkun, G., and Horvitz, H. R. (1988). The *C. elegans* cell lineage and differentiation gene *unc-86* encodes a protein with a homeodomain and extended similarity to transcription factors. *Cell* **55,** 757–769.

Fukushige, T., Siddiqui, Z. K., Chou, M., Culotti, J. G., Gogonea, C. B., Siddiqui, S. S., and Hamelin, M. (1999). MEC-12, an α-tubulin required for touch sensitivity in *C. elegans*. *J. Cell. Sci.* **112**(Pt 3), 395–403.

García-Añoveros, J., Garcia, J. A., Liu, J. D., and Corey, D. P. (1998). The nematode degenerin UNC-105 forms ion channels that are activated by degeneration- or hypercontraction-causing mutations. *Neuron* **20,** 1231–1241.

García-Añoveros, J., Ma, C., and Chalfie, M. (1995). Regulation of *Caenorhabditis elegans* degenerin proteins by a putative extracellular domain. *Curr. Biol.* **5,** 441–448.

Gong, Z., Son, W., Chung, Y. D., Kim, J., Shin, D. W., McClung, C. A., Lee, Y., Lee, H. W., Chang, D. J., Kaang, B. K., Cho, H., Oh, U., Hirsh, J., Kernan, M. J., and Kim, C. (2004). Two interdependent TRPV channel subunits, inactive and Nanchung, mediate hearing in Drosophila. *J. Neurosci.* **24,** 9059–9066.

Goodman, M. B., Ernstrom, G. G., Chelur, D. S., O' Hagan, R., Yao, C. A., and Chalfie, M. (2002). MEC-2 regulates *C. elegans* DEG/ENaC channels needed for mechanosensation. *Nature* **415,** 1039–1042.

Goodman, M. B., Hall, D. H., Avery, L., and Lockery, S. R. (1998). Active currents regulate sensitivity and dynamic range in *C. elegans* neurons. *Neuron* **20,** 763–772.

Goodman, M. B., and Schwarz, E. M. (2003). Transducing Touch in *Caenorhabditis elegans*. *Annu. Rev. Physiol.* **65,** 429–452.

Gu, G., Caldwell, G. A., and Chalfie, M. (1996). Genetic interactions affecting touch sensitivity in *Caenorhabditis elegans*. *Proc. Natl. Acad. Sci. USA* **93,** 6577–6582.

Hamill, O. P., and McBride, D. W., Jr. (1996). The pharmacology of mechanogated membrane ion channels. *Pharmacol. Rev.* **48,** 231–252.

Hanaoka, K., Qian, F., Boletta, A., Bhunia, A. K., Piontek, K., Tsiokas, L., Sukhatme, V. P., Guggino, W. B., and Germino, G. G. (2000). Co-assembly of polycystin-1 and -2 produces unique cation-permeable currents. *Nature* **408,** 990–994.

Hardie, R. C., and Minke, B. (1992). The *trp* gene is essential for a light-activated Ca^{2+} channel in Drosophila photoreceptors. *Neuron* **8,** 643–651.

Hart, A. C., Kass, J., Shapiro, J. E., and Kaplan, J. M. (1999). Distinct signaling pathways mediate touch and osmosensory responses in a polymodal sensory neuron. *J. Neurosci.* **19,** 1952–1958.

Hart, A. C., Sims, S., and Kaplan, J. M. (1995). Synaptic code for sensory modalities revealed by *C. elegans* GLR-1 glutamate receptor. *Nature* **378,** 82–85.

Hilliard, M. A., Apicella, A. J., Kerr, R., Suzuki, H., Bazzicalupo, P., and Schafer, W. R. (2005). *In vivo* imaging of *C. elegans* ASH neurons: Cellular response and adaptation to chemical repellents. *EMBO J.* **24,** 63–72.

Hodgkin, J. (1983). Male phenotypes and mating efficiency in *C. elegans*. *Genetics* **103,** 43–64.

Hong, K., and Driscoll, M. (1994). A transmembrane domain of the putative channel subunit MEC-4 influences mechanotransduction and neurodegeneration in *C. elegans*. *Nature* **367,** 470–473.

Hong, K., Mano, I., and Driscoll, M. (2000). *In vivo* structure-function analyses of *Caenorhabditis elegans* MEC-4, a candidate mechanosensory ion channel subunit. *J. Neurosci.* **20,** 2575–2588.

Howard, J., and Hudspeth, A. J. (1988). Compliance of the hair bundle associated with gating of mechanoelectrical transduction channels in the bullfrog's saccular hair cell. *Neuron* **1,** 189–199.

Huang, M., and Chalfie, M. (1994). Gene interactions affecting mechanosensory transduction in *Caenorhabditis elegans*. *Nature* **367,** 467–469.

Huang, M., Gu, G., Ferguson, E. L., and Chalfie, M. (1995). A stomatin-like protein necessary for mechanosensation in *C. elegans*. *Nature* **378,** 292–295.

Hudspeth, A. J. (1989). How the ear's works work. *Nature* **341,** 397–404.

Jan, Y. N., and Jan, L. Y. (1993). The peripheral nervous system. *In* "The Development of *Drosophila melanogaster*" (M. Bate and A. M. Arias, Eds.), pp. 1207–1244. Cold Spring Harbor Laboratory Press, Plainview, New York.

Jauregui, A. R., and Barr, M. M. (2005). Functional characterization of the *C. elegans* nephrocystins NPHP-1 and NPHP-4 and their role in cilia and male sensory behaviors. *Exp. Cel. Res.* **305,** 333–342.

Jospin, M., Mariol, M. C., Segalat, L., and Allard, B. (2004). Patch clamp study of the UNC-105 degenerin and its interaction with the LET-2 collagen in *Caenorhabditis elegans* muscle. *J. Physiol.* **557,** 379–388.

Kahn-Kirby, A. H., Dantzker, J. L., Apicella, A. J., Schafer, W. R., Browse, J., Bargmann, C. I., and Watts, J. L. (2004). Specific polyunsaturated fatty acids drive TRPV-dependent sensory signaling *in vivo*. *Cell* **119,** 889–900.

Kaplan, J. M., and Horvitz, H. R. (1993). A dual mechanosensory and chemosensory neuron in *Caenorhabditis elegans*. *Proc. Natl. Acad. Sci. USA* **90,** 2227–2231.

Kellenberger, S., and Schild, L. (2002). Epithelial sodium channel/degenerin family of ion channels: A variety of functions for a shared structure. *Physiol. Rev.* **82,** 735–767.

Kernan, M., Cowan, D., and Zuker, C. (1994). Genetic dissection of mechanosensory transduction: Mechanoreception-defective mutations of *Drosophila*. *Neuron* **12,** 1195–1206.

Krishtal, O. (2003). The ASICs: Signaling molecules? Modulators? *Trends Neurosci.* **26,** 477–483.

Lee, R. Y., Sawin, E. R., Chalfie, M., Horvitz, H. R., and Avery, L. (1999). EAT-4, a homolog of a mammalian sodium-dependent inorganic phosphate cotransporter, is necessary for glutamatergic neurotransmission in *Caenorhabditis elegans*. *J. Neurosci.* **19,** 159–167.

Liedtke, W., Tobin, D. M., Bargmann, C. I., and Friedman, J. M. (2003). Mammalian TRPV4 (VR-OAC) directs behavioral responses to osmotic and mechanical stimuli in *Caenorhabditis elegans*. *Proc. Natl. Acad. Sci. USA* **100**(Suppl. 2), 14531–14536.

Liu, D. W., and Thomas, J. H. (1994). Regulation of a periodic motor program in *C. elegans*. *J. Neurosci.* **14,** 1953–1962.

Liu, K. S., and Sternberg, P. W. (1995). Sensory regulation of male mating behavior in *Caenorhabditis elegans*. *Neuron* **14,** 79–89.

Lockery, S. R., and Goodman, M. B. (1998). Tight-seal whole-cell patch clamping of *Caenorhabditis elegans* neurons. *Meth. Enzymol.* **293,** 201–217.

Loer, C. M., and Kenyon, C. J. (1993). Serotonin-deficient mutants and male mating behavior in the nematode *Caenorhabditis elegans*. *J. Neurosci.* **13,** 5407–5417.

Lundquist, E. A., Herman, R. K., Rogalski, T. M., Mullen, G. P., Moerman, D. G., and Shaw, J. E. (1996). The *mec-8* gene of *C. elegans* encodes a protein with two RNA recognition motifs and regulates alternative splicing of *unc-52* transcripts. *Development* **122,** 1601–1610.

Maricq, A. V., Peckol, E., Driscoll, M., and Bargmann, C. I. (1995). Mechanosensory signaling in *C. elegans* mediated by the GLR-1 glutamate receptor. *Nature* **378,** 78–81.

McIntire, S. L., Jorgensen, E., Kaplan, J., and Horvitz, H. R. (1993). The GABAergic nervous system of *Caenorhabditis elegans*. *Nature* **364,** 337–341.

Mellem, J. E., Brockie, P. J., Zheng, Y., Madsen, D. M., and Maricq, A. V. (2002). Decoding of polymodal sensory stimuli by postsynaptic glutamate receptors in *C. elegans*. *Neuron* **36,** 933–944.

Mitani, S., Du, H., Hall, D. H., Driscoll, M., and Chalfie, M. (1993). Combinatorial control of touch receptor neuron expression in *Caenorhabditis elegans*. *Development* **119,** 773–783.

Montell, C. (2003). The venerable inveterate invertebrate TRP channels. *Cell. Calcium* **33,** 409–417.

Nagata, K., Duggan, A., Kumar, G., and García-Añoveros, J. (2005). Nociceptor and hair cell transducer properties of TRPA1, a channel for pain and hearing. *J. Neurosci.* **25,** 4052–4061.

Nauli, S. M., Alenghat, F. J., Luo, Y., Williams, E., Vassilev, P., Li, X., Elia, A. E., Lu, W., Brown, E. M., Quinn, S. J., Ingber, D. E., and Zhou, J. (2003). Polycystins 1 and 2 mediate mechanosensation in the primary cilium of kidney cells. *Nat. Genet.* **33,** 129–137.

Nauli, S. M., and Zhou, J. (2004). Polycystins and mechanosensation in renal and nodal cilia. *Bioessays* **26,** 844–856.

O'Hagan, R., Chalfie, M., and Goodman, M. B. (2005). The MEC-4 DEG/ENaC channel of *Caenorhabditis elegans* touch receptor neurons transduces mechanical signals. *Nat. Neurosci.* **8,** 43–50.

Patel, A. J., Honore, E., Maingret, F., Lesage, F., Fink, M., Duprat, F., and Lazdunski, M. (1998). A mammalian two pore domain mechano-gated S-like K^+ channel. *EMBO J.* **17,** 4283–4290.

Peden, E. M., and Barr, M. M. (2005). The KLP-6 kinesin is required for male mating behaviors and polycystin localization in *Caenorhabditis elegans*. *Curr. Biol.* **15,** 394–404.

Perkins, L. A., Hedgecock, E. M., Thomson, J. N., and Culotti, J. G. (1986). Mutant sensory cilia in the nematode *Caenorhabditis elegans*. *Dev. Biol.* **117,** 456–487.

Portman, D. S., and Emmons, S. W. (2004). Identification of *C. elegans* sensory ray genes using whole-genome expression profiling. *Dev. Biol.* **270,** 499–512.

Qian, F., Germino, F. J., Cai, Y., Zhang, X., Somlo, S., and Germino, G. G. (1997). PKD1 interacts with PKD2 through a probable coiled-coil domain. *Nat. Genet.* **16,** 179–183.

Rajaram, S., Spangler, T. L., Sedensky, M. M., and Morgan, P. G. (1999). A stomatin and a degenerin interact to control anesthetic sensitivity in *Caenorhabditis elegans*. *Genetics* **153,** 1673–1682.

Rankin, C. H. (2000). Context conditioning in habituation in the nematode *Caenorhabditis elegans*. *Behav. Neurosci.* **114,** 496–505.

Rankin, C. H., Beck, C. D., and Chiba, C. M. (1990). *Caenorhabditis elegans*: A new model system for the study of learning and memory. *Behav. Brain Res.* **37,** 89–92.

Rankin, C. H., and Broster, B. S. (1992). Factors affecting habituation and recovery from habituation in the nematode *Caenorhabditis elegans*. *Behav. Neurosci.* **106,** 239–249.

Rankin, C. H., and Wicks, S. R. (2000). Mutations of the *Caenorhabditis elegans* brain-specific inorganic phosphate transporter *eat-4* affect habituation of the tap-withdrawal response without affecting the response itself. *J. Neurosci.* **20,** 4337–4344.

Rose, J. K., Kaun, K. R., Chen, S. H., and Rankin, C. H. (2003). GLR-1, a non-NMDA glutamate receptor homolog, is critical for long-term memory in *Caenorhabditis elegans*. *J. Neurosci.* **23,** 9595–9599.

Rose, J. K., and Rankin, C. H. (2001). Analyses of habituation in *Caenorhabditis elegans*. *Learn. Mem.* **8,** 63–69.

Sanyal, S., Wintle, R. F., Kindt, K. S., Nuttley, W. M., Arvan, R., Fitzmaurice, P., Bigras, E., Merz, D. C., Hebert, T. E., van der Kooy, D., Schafer, W. R., Culotti, J. G., and Van Tol, H. H. (2004). Dopamine modulates the plasticity of mechanosensory responses in *Caenorhabditis elegans*. *EMBO J.* **23,** 473–482.

Savage, C., Hamelin, M., Culotti, J. G., Coulson, A., Albertson, D. G., and Chalfie, M. (1989). *mec-7* is a β-tubulin gene required for the production of 15-protofilament microtubules in *Caenorhabditis elegans*. *Genes Dev.* **3,** 870–881.

Savage, C., Xue, Y., Mitani, S., Hall, D., Zakhary, R., and Chalfie, M. (1994). Mutations in the *Caenorhabditis elegans* β-tubulin gene *mec-7*: Effects on microtubule assembly and stability and on tubulin autoregulation. *J. Cell Sci.* **107**(Pt 8), 2165–2175.

Sawin, E. R., Ranganathan, R., and Horvitz, H. R. (2000). *C. elegans* locomotory rate is modulated by the environment through a dopaminergic pathway and by experience through a serotonergic pathway. *Neuron* **26,** 619–631.

Sedensky, M. M., Siefker, J. M., Koh, J. Y., Miller, D. M., 3rd, and Morgan, P. G. (2004). A stomatin and a degenerin interact in lipid rafts of the nervous system of *Caenorhabditis elegans*. *Am. J. Physiol. Cell. Physiol.* **287,** C468–C474.

Sedensky, M. M., Siefker, J. M., and Morgan, P. G. (2001). Model organisms: New insights into ion channel and transporter function. Stomatin homologues interact in *Caenorhabditis elegans*. *Am. J. Physiol. Cell. Physiol.* **280,** C1340–C1348.

Shreffler, W., Magardino, T., Shekdar, K., and Wolinsky, E. (1995). The *unc-8* and *sup-40* genes regulate ion channel function in *Caenorhabditis elegans* motor neurons. *Genetics* **139,** 1261–1272.

Sidi, S., Friedrich, R. W., and Nicolson, T. (2003). NompC TRP channel required for vertebrate sensory hair cell mechanotransduction. *Science* **301,** 96–99.

Snyder, P. M., Bucher, D. B., and Olson, D. R. (2000). Gating induces a conformational change in the outer vestibule of ENaC. *J. Gen. Physiol.* **116,** 781–790.

Starich, T. A., Herman, R. K., Kari, C. K., Yeh, W. H., Schackwitz, W. S., Schuyler, M. W., Collet, J., Thomas, J. H., and Riddle, D. L. (1995). Mutations affecting the chemosensory neurons of *Caenorhabditis elegans*. *Genetics* **139,** 171–188.

Stewart, G. W. (1993). The membrane defect in hereditary stomatocytosis. *Baillieres Clin. Haematol.* **6,** 371–399.

Sulston, J. E., and Horvitz, H. R. (1977). Postembryonic cell lineages of the nematode, *Caenorhabditis elegans*. *Dev. Biol.* **56,** 110–156.

Sulston, J., Dew, M., and Brenner, S. (1975). Dopaminergic neurons in the nematode *Caenorhabditis elegans*. *J. Comp. Neurol.* **163,** 215–226.

Sulston, J. E., Albertson, D. G., and Thomson, J. N. (1980). The *Caenorhabditis elegans* male: Postembryonic development of nongonadal structures. *Dev. Biol.* **78,** 542–576.

Suzuki, H., Kerr, R., Bianchi, L., Frokjaer-Jensen, C., Slone, D., Xue, J., Gerstbrein, B., Driscoll, M., and Schafer, W. R. (2003). *In vivo* imaging of *C. elegans* mechanosensory neurons demonstrates a specific role for the MEC-4 channel in the process of gentle touch sensation. *Neuron* **39,** 1005–1017.

Tabish, M., Siddiqui, Z. K., Nishikawa, K., and Siddiqui, S. S. (1995). Exclusive expression of *C. elegans osm-3* kinesin gene in chemosensory neurons open to the external environment. *J. Mol. Biol.* **247,** 377–389.

Tavernarakis, N., and Driscoll, M. (2000). *Caenorhabditis elegans* degenerins and vertebrate ENaC ion channels contain an extracellular domain related to venom neurotoxins. *J. Neurogenet.* **13,** 257–264.

Tavernarakis, N., Shreffler, W., Wang, S., and Driscoll, M. (1997). *unc-8*, a DEG/ENaC family member, encodes a subunit of a candidate mechanically gated channel that modulates *C. elegans* locomotion. *Neuron* **18,** 107–119.

Tobin, D., Madsen, D., Kahn-Kirby, A., Peckol, E., Moulder, G., Barstead, R., Maricq, A., and Bargmann, C. (2002). Combinatorial expression of TRPV channel proteins defines their sensory functions and subcellular localization in *C. elegans* neurons. *Neuron* **35,** 307–318.

Toker, A. S., Teng, Y., Ferreira, H. B., Emmons, S. W., and Chalfie, M. (2003). The *Caenorhabditis elegans spalt*-like gene *sem-4* restricts touch cell fate by repressing the selector Hox gene *egl-5* and the effector gene *mec-3*. *Development* **130,** 3831–3840.

Tracey, W. D., Jr., Wilson, R. I., Laurent, G., and Benzer, S. (2003). *Painless*, a Drosophila gene essential for nociception. *Cell* **113,** 261–273.

Troemel, E. R., Chou, J. H., Dwyer, N. D., Colbert, H. A., and Bargmann, C. I. (1995). Divergent seven transmembrane receptors are candidate chemosensory receptors in *C. elegans*. *Cell* **83,** 207–218.

Tsalik, E. L., Niacaris, T., Wenick, A. S., Pau, K., Avery, L., and Hobert, O. (2003). LIM homeobox gene-dependent expression of biogenic amine receptors in restricted regions of the *C. elegans* nervous system. *Dev. Biol.* **263,** 81–102.

Tsiokas, L., Kim, E., Arnould, T., Sukhatme, V. P., and Walz, G. (1997). Homo- and heterodimeric interactions between the gene products of PKD1 and PKD2. *Proc. Natl. Acad. Sci. USA* **94,** 6965–6970.

Vogel, B. E., and Hedgecock, E. M. (2001). Hemicentin, a conserved extracellular member of the immunoglobulin superfamily, organizes epithelial and other cell attachments into oriented line-shaped junctions. *Development* **128,** 883–894.

Walker, R. G., Willingham, A. T., and Zuker, C. S. (2000). A Drosophila mechanosensory transduction channel. *Science* **287,** 2229–2234.

Walthall, W. W., and Chalfie, M. (1988). Cell-cell interactions in the guidance of late-developing neurons in *Caenorhabditis elegans*. *Science* **239,** 643–645.

Ward, S., Thomson, N., White, J. G., and Brenner, S. (1975). Electron microscopical reconstruction of the anterior sensory anatomy of the nematode *Caenorhabditis elegans*. *J. Comp. Neurol.* **160,** 313–337.

Way, J. C., and Chalfie, M. (1988). *mec-3*, a homeobox-containing gene that specifies differentiation of the touch receptor neurons in *C. elegans*. *Cell* **54,** 5–16.

Way, J. C., and Chalfie, M. (1989). The *mec-3* gene of *Caenorhabditis elegans* requires its own product for maintained expression and is expressed in three neuronal cell types. *Genes Dev.* **3,** 1823–1833.

White, J. G., Southgate, E., Thomson, J. N., and Brenner, S. (1986). The Structure of the nervous system of the nematode *Caenorhabditis elegans*. *Phil. Trans. Roy. Soc. Lond.* **314,** 1–340.

Wicks, S. R., de Vries, C. J., van Luenen, H. G., and Plasterk, R. H. (2000). CHE-3, a cytosolic dynein heavy chain, is required for sensory cilia structure and function in *Caenorhabditis elegans*. *Dev. Biol.* **221,** 295–307.

Wicks, S. R., and Rankin, C. H. (1995). Integration of mechanosensory stimuli in *Caenorhabditis elegans*. *J. Neurosci.* **15,** 2434–2444.

Wu, J., Duggan, A., and Chalfie, M. (2001). Inhibition of touch cell fate by *egl-44* and *egl-46* in *C. elegans*. *Genes Dev.* **15,** 789–802.

Xue, D., Finney, M., Ruvkun, G., and Chalfie, M. (1992). Regulation of the *mec-3* gene by the *C. elegans* homeoproteins UNC-86 and MEC-3. *EMBO J.* **11,** 4969–4979.

Xue, D., Tu, Y., and Chalfie, M. (1993). Cooperative interactions between the *Caenorhabditis elegans* homeoproteins UNC-86 and MEC-3. *Science* **261,** 1324–1328.

Yochem, J., Bell, L. R., and Herman, R. K. (2004). The identities of *sym-2, sym-3* and *sym-4,* three genes that are synthetically lethal with *mec-8* in *Caenorhabditis elegans. Genetics* **168,** 1293–1306.

Zhang, S. (2004). "Stomatin Gene Family in *Caenorhabditis elegans*" PhD Thesis. Columbia University, New York City, NY.

Zhang, S., Arnadottir, J., Keller, C., Caldwell, G. A., Yao, C. A., and Chalfie, M. (2004). MEC-2 is recruited to the putative mechanosensory complex in *C. elegans* touch receptor neurons through its stomatin-like domain. *Curr. Biol.* **14,** 1888–1896.

Zhang, Y., Ma, C., Delohery, T., Nasipak, B., Foat, B. C., Bounoutas, A., Bussemaker, H. J., Kim, S. K., and Chalfie, M. (2002). Identification of genes expressed in *C. elegans* touch receptor neurons. *Nature* **418,** 331–335.

INDEX

CONTENTS OF RECENT VOLUMES

Volume 39

Volume 40

Volume 41

Volume 42

Volume 43

Volume 44

Volume 45

Volume 46

Volume 47

Volume 48

Volume 49

Volume 50

Volume 51

Volume 52

Volume 53

Volume 54

Volume 55

Volume 56

Volume 57

Volume 58

Volume 59

Volume 60

Volume 61

Volume 62

Volume 63

Volume 64

Volume 65

Volume 66

Volume 67

Volume 68

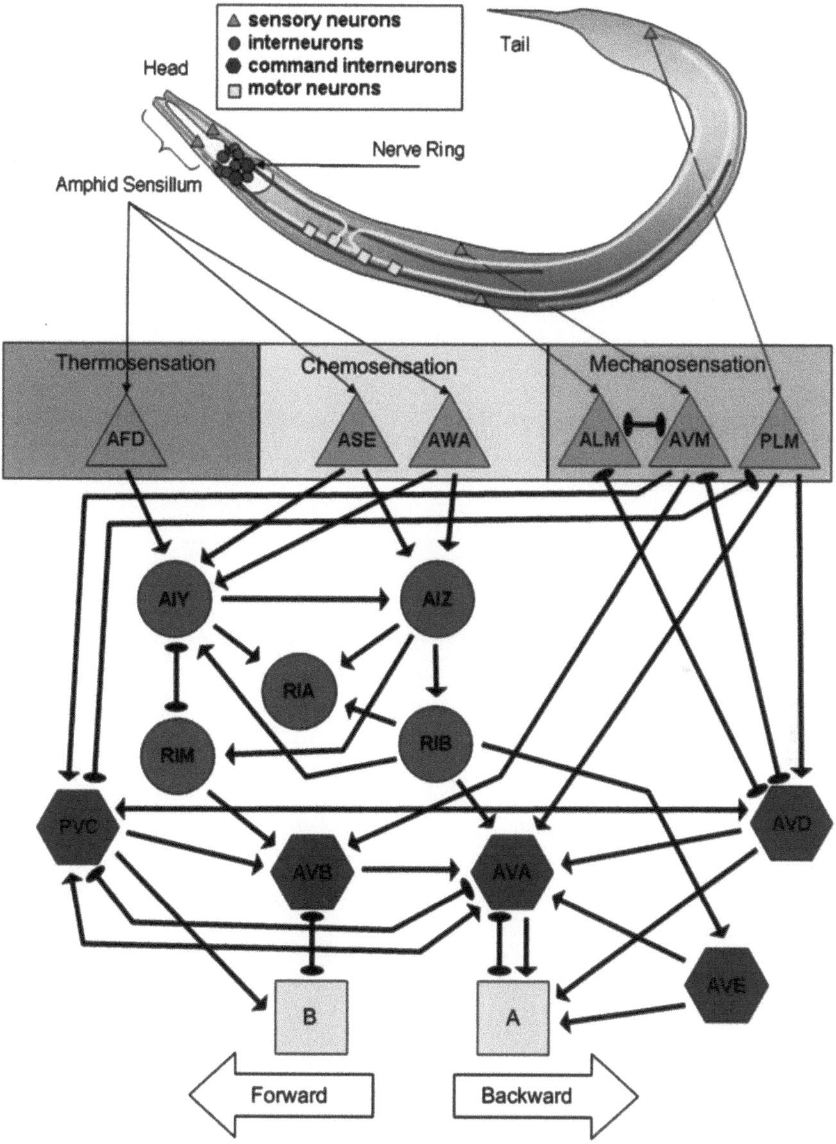

GILES *ET AL.*, FIG. 1. Simplified and integrated schematic diagram of the major elements of the neural circuits involved in mechanosensory, chemosensory, and thermosensory behaviors that have been used to demonstrate learning and memory and their anatomical locations within the worm. Green triangles represent sensory neurons, red circles represent interneurons, blue hexagons represent command interneurons, and yellow squares represent pools of motor neurons. Arrows represent chemical synaptic connections while bars represent gap junctions. Based on the hypothesized neural circuits created by Chalfie *et al.* (1985), Wicks and Rankin (1995), Mori and Oshima (1995), and Hobert (2003).

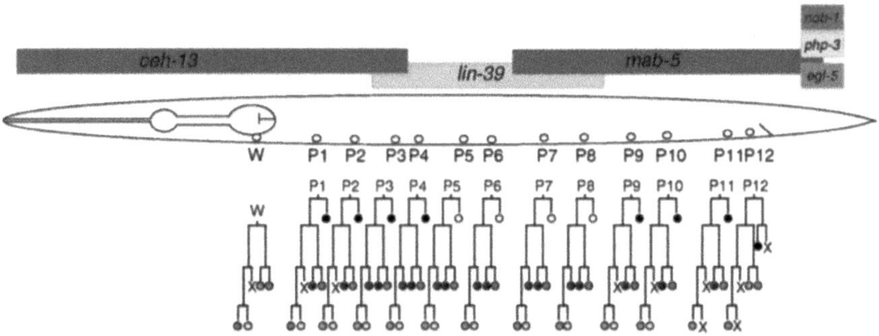

Aamodt and Aamodt, Fig. 1. The HOM-C genes *ceh-13*, *lin-39*, *mab-5*, *egl-5*, *php-3*, and *nob-1* affect the fate of neurons and other cells along the length of *C. elegans*. The colored boxes represent the approximate regions over which each HOM-C gene is expressed. The worm depicted is at the early first larval (L1) stage. The circles labeled W, P1, P2, and so on represent the P-neuroblasts that undergo the stereotyped cell divisions shown below. While each of the P-neuroblasts undergo a similar set of divisions, there are variations on this theme along the length of the animal and there are differences in the fates of the neurons generated by these lineages. The cell fates are indicated by colored circles as follows: orange are AVF neurons, yellow are VB neurons, green are VA neurons, blue are VD neurons, purple are AS neurons, red are VC neurons, and gray is a PDB neuron. The black cells form hypodermis and the white cells form vulva. Some of the hypodermal cells and all of the vulval cells undergo further cell divisions at the L3 stage that are not shown.

A

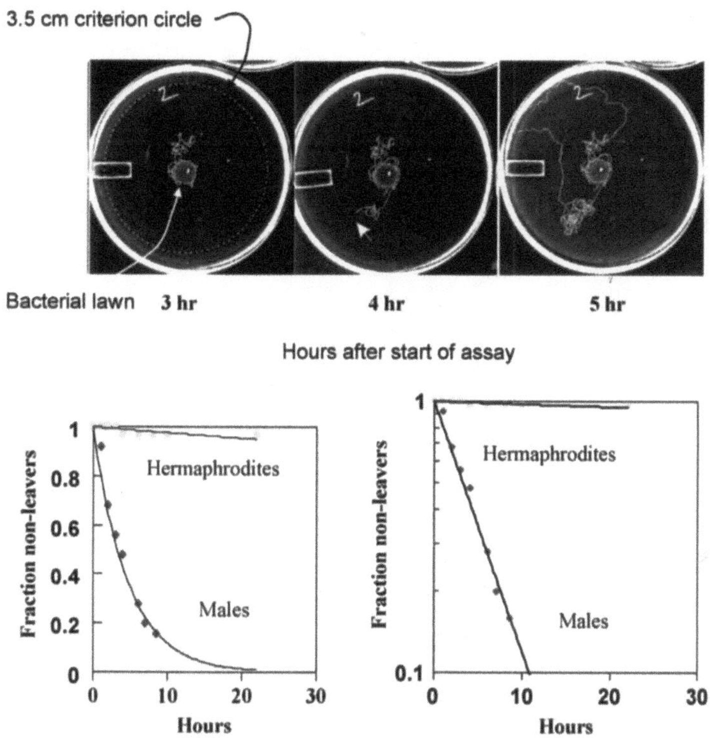

EMMONS, FIG. 1. Males leave food in a mate-searching assay. A. A plate in an assay for mate-searching ("leaving" assay) is shown at three times after setting up the assay. At time 0, a single male was placed on a 1 cm bacterial lawn on agar in the center of a 9 cm petri dish. As the male makes excursions off the lawn he leaves an indentation on the agar surface that can be photographed as a wiggly white line. During the first 3 h, the male made an excursion off the lawn and returned to it. At 4 h, the male was at the position indicated by the arrowhead. Between 4 and 5 h the male track traverses a 3.5 cm radius criterion circle and the male is scored as a leaver. B. Typical plots for 20 worms in a leaving assay of the fraction of worms that are non-leavers. Males (lower curves) leave following exponential kinetics (straight line on a semi log plot, right graph) while hermaphrodites remain on the food (upper curves). Part A is reproduced from Lipton *et al.*, 2004. Copyright 2004 by the Society of Neuroscience.

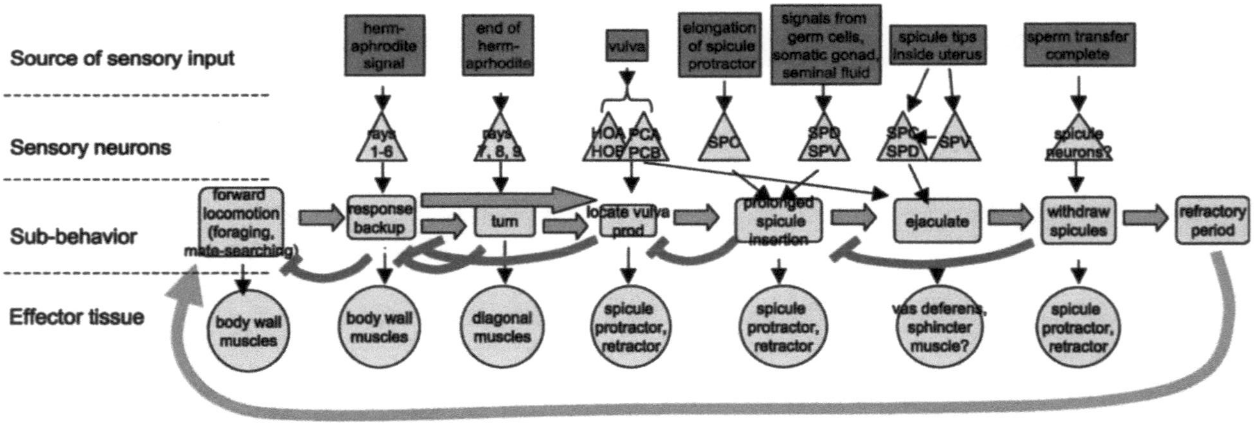

EMMONS, FIG. 3. Copulation proceeds through a series of sub-behaviors. Each sub-behavior is initiated and reinforced by sensory input through identified sensory neurons. Although the sources of signals are shown, none of the molecular or mechanical cues detected by sensory neurons have been identified to date. The most prominent sensory input associated with each sub-behavior is shown; some sub-behaviors can be sustained by alternative input. For example, the hook (HOA, HOB) and post-cloacal sensilla (PCA, PCB, PCC) can each sustain the response and back up sub-behaviors. Each sub-behavior tends to lead the male to a new set of inputs, which causes a switch in behavior, the ongoing sub-behavior is suppressed, and a new sub-behavior is initiated. The pathway need not occur in fixed order. It can jump over steps or back up in the sequence as necessary.

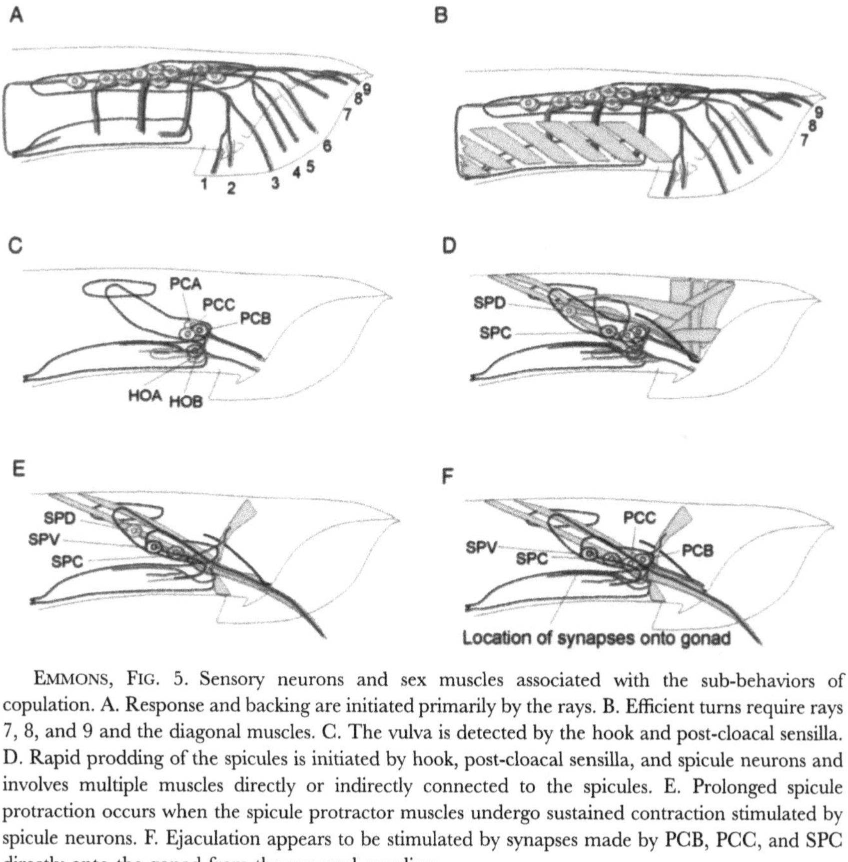

EMMONS, FIG. 5. Sensory neurons and sex muscles associated with the sub-behaviors of copulation. A. Response and backing are initiated primarily by the rays. B. Efficient turns require rays 7, 8, and 9 and the diagonal muscles. C. The vulva is detected by the hook and post-cloacal sensilla. D. Rapid prodding of the spicules is initiated by hook, post-cloacal sensilla, and spicule neurons and involves multiple muscles directly or indirectly connected to the spicules. E. Prolonged spicule protraction occurs when the spicule protractor muscles undergo sustained contraction stimulated by spicule neurons. F. Ejaculation appears to be stimulated by synapses made by PCB, PCC, and SPC directly onto the gonad from the pre-anal ganglion.

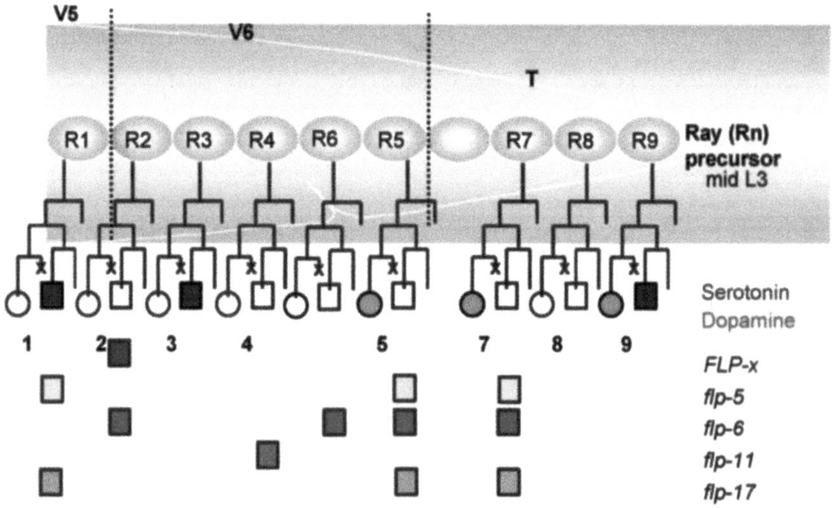

EMMONS, FIG. 6. Neurotransmitters expressed by the rays. Ray precursor cells (Rn) in the posterior male seam express the ray sublineage during the late L3 early L4 larval stages as shown. Rn.aaa: A-type neuron (circle); Rn.aap: programmed cell death; Rn.apa: B-type neuron (square); Rn.app: structural cell of the ray; Rn.p: hypodermal cell. The *flp* genes encode FMRFamide peptide neurotransmitters. Adapted from Lints *et al.* (2004).

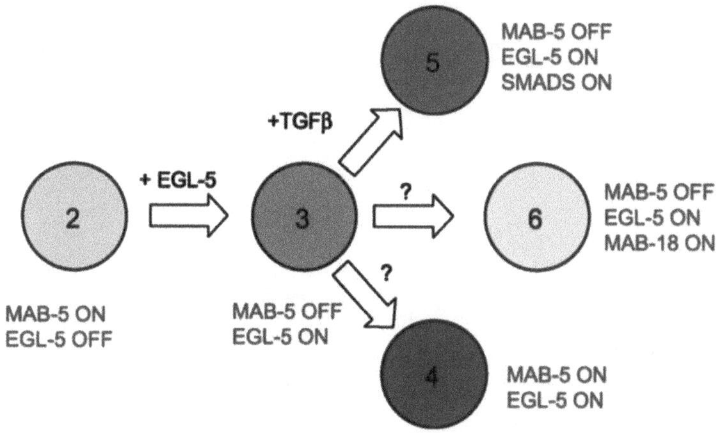

EMMONS, FIG. 7. A transcription factor code establishes the identities of the V6 rays (the rays descended from the L1 blast cell V6). This schematic diagram shows the hierarchical relationships among the identities of the V6 rays. The differing characteristics of the V6 rays, rays 2 to 6, are determined in part by the Hox genes *mab-5* and *egl-5*, the Pax6 gene *mab-18*, and a TGFβ signal. In *egl-5* mutants, ray 6 is not made because of a lineage transformation while the remaining rays assume the characteristics of ray 2, which may be considered a ground state. The TGFβ signal is selectively received by ray 5. Further signals involved in establishing the identities of rays 4 and 6 remain to be identified. Reprinted from Lints *et al.* (2004), with permission from Elsevier.

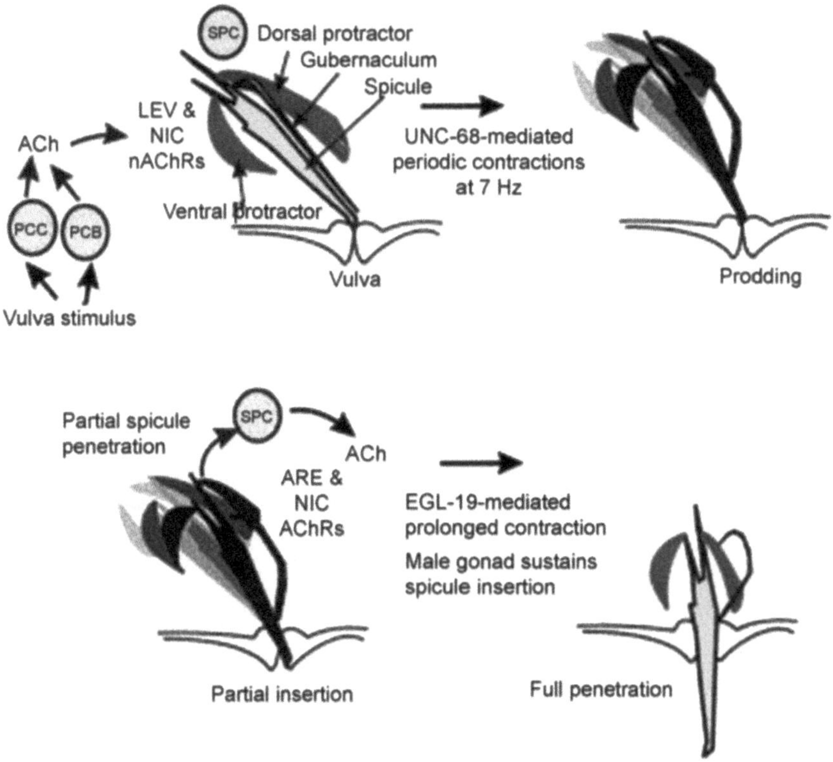

EMMONS, FIG. 8. The roles of two cholinergic pathways in prodding and penetration behavior of the spicules. Top. Stimulated by ACh from post-cloacal sensilla neurons PCC and PCB and spicule neuron SPC, the spicule protractor muscle twitches, causing the spicules to rapidly prod the region of the vulva. Prodding involves release of intracellular Ca^{++} by the ryanodine-sensitive, sarcoplasmic Ca^{++} channel UNC-68. Bottom. When the spicules penetrate the vulva, SPC, which has a presumptively proprioceptive dendrite attached to the spicule protractor muscle, releases ACh and stimulates a prolonged contraction. Prolonged protraction involves entry of extracellular Ca^{++} through the L-type voltage-gated channel EGL-19 and is sustained by substances in seminal fluid. Reprinted from Garcia *et al.* (2001), with permission from Elsevier.

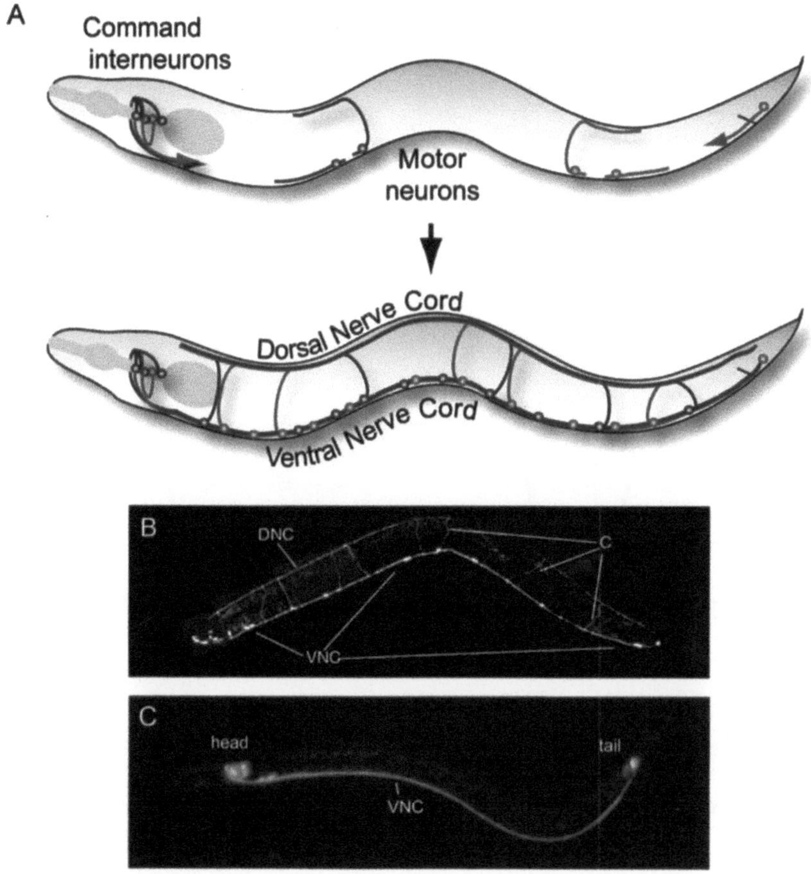

VON STETINA *ET AL.*, FIG. 1. The *C. elegans* motor circuit. A. Command interneurons in the head and tail extend processes into the ventral nerve cord to synapse with specific classes of motor neurons. "Red" and "Blue" circuits denote separate sets of command interneurons and motor neuron partners. Selected motor neurons shown here are presynaptic to adjacent body muscle cells in either the ventral or dorsal nerve cords. B. GFP driven by the *unc-47* promoter labels the GABA circuit with motor neurons extending processes in the ventral nerve cord (VNC) and to the dorsal nerve cord (DNC) via commissures (C) (reprinted from Francis, 2003). C. A subset of neurons, including the command interneurons AVA, AVD, AVE (head) and PVC (tail) are marked with the *nmr-1*::GFP reporter to show projections into the VNC (adapted from Francis, 2003).

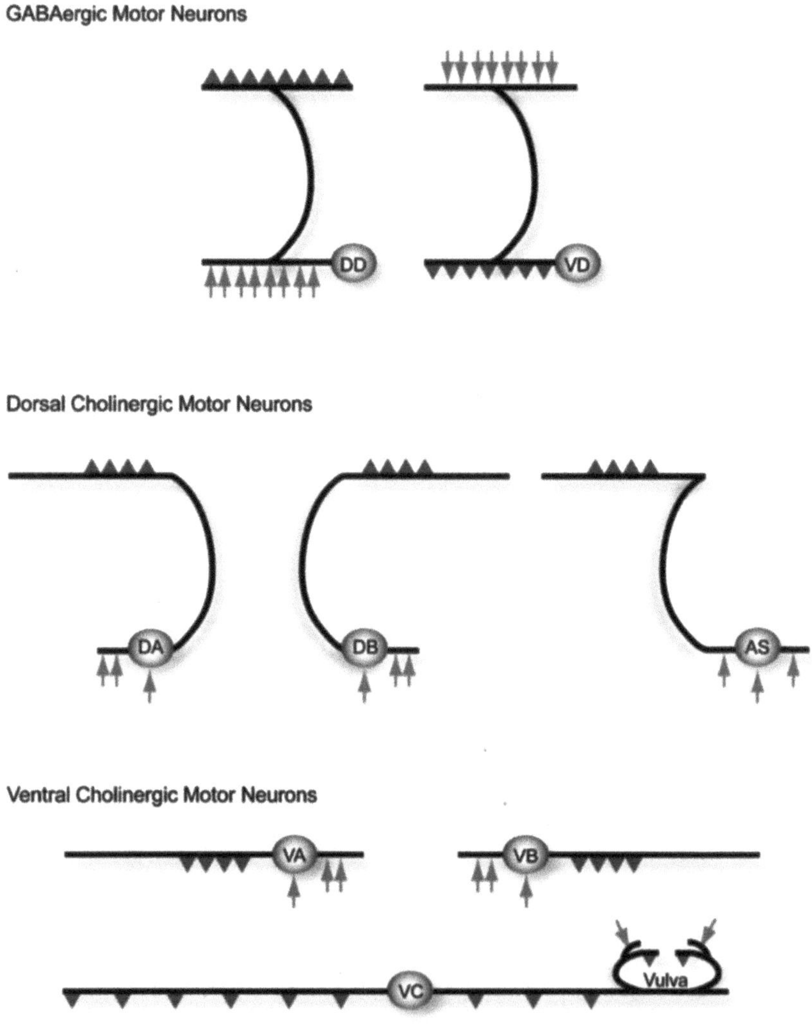

VON STETINA *ET AL.*, FIG. 4. Morphology of adult ventral cord motor neurons. Representative members of each motor neuron class are shown here. Note that five of the motor neuron classes (DA, DB, DD, AS, VD) extend commissures to the DNC whereas three (VA, VB, VC) are restricted to the VNC. Each motor neuron is representative of its class although minor differences in morphology do occur within each class. For example, in contrast to other VC class motor neurons, VC4 and VC5 do not make NMJs with body wall muscle as indicated in this diagram. Green arrows depict input from other neurons. Red triangles denote NMJs with body wall muscle cells.

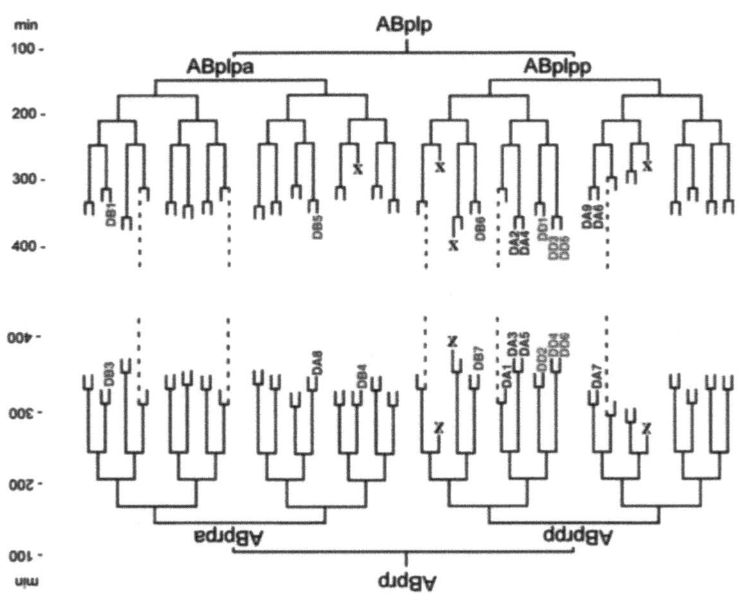

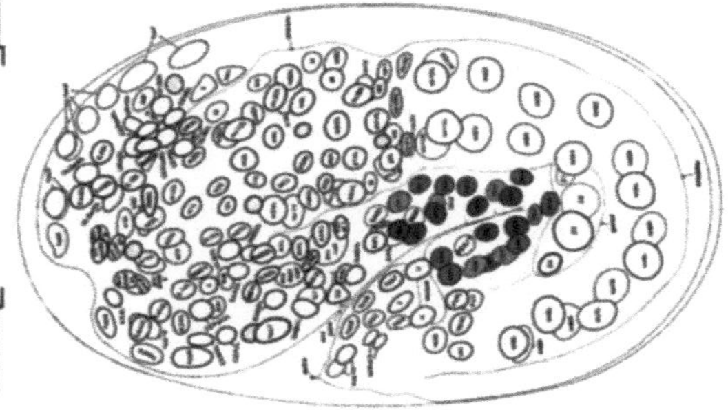

Von Stetina *et al.*, Fig. 5. Embryonic ventral cord motor neurons are derived from specific cell lineages. A. All but one of the embryonic VNC motor neurons are derived from two embryonic precursor cells, ABplp and ABprp (DB2 arises from ABa). The ABprp lineage is inverted to feature the overall bilateral symmetry of these cell divisions. The Y-axis refers to minutes after fertilization (at 20°C). (Adapted from Hallam *et al.*, 2000.) B. Diagram of medial/left hand side of comma stage embryo to demonstrate the placement of the embryonic motor neurons along the ventral surface. (adapted from Sulston *et al.*, 1983.)

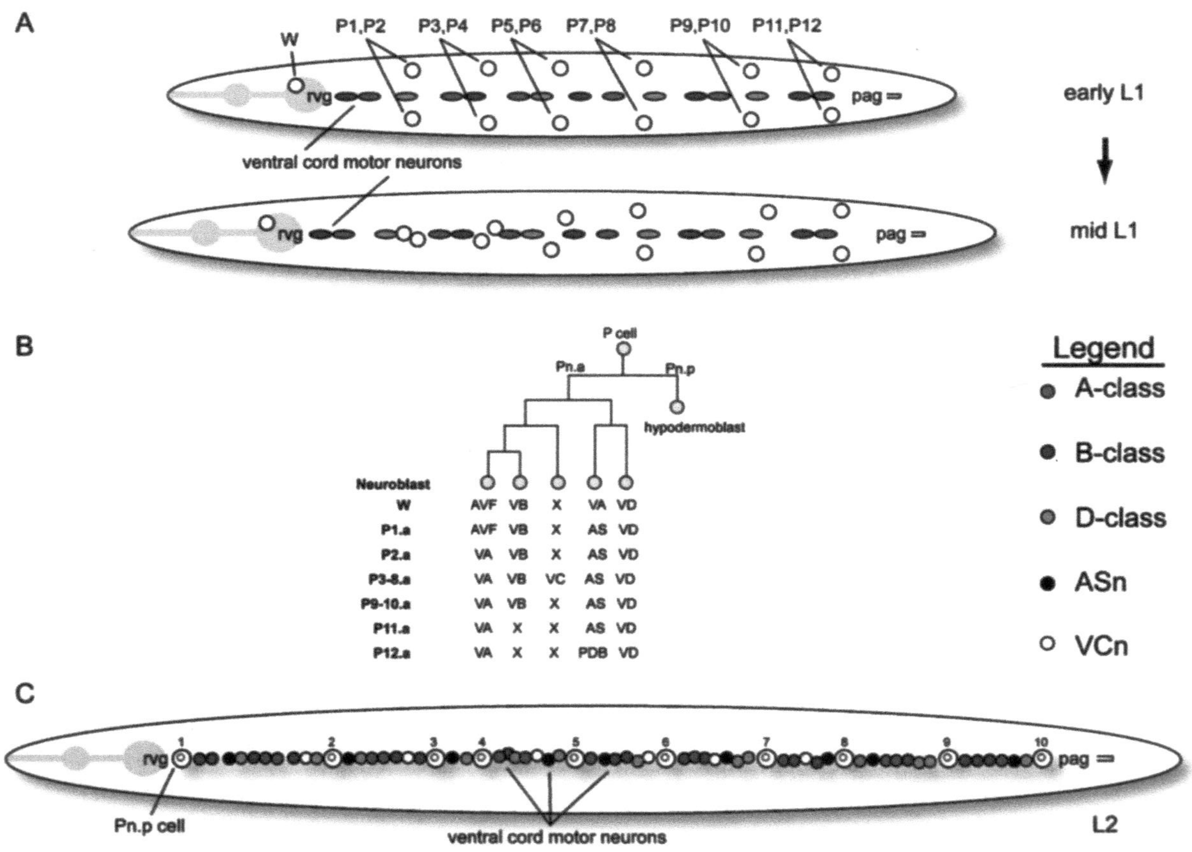
A
W
P1,P2
P3,P4
P5,P6
P7,P8
P9,P10
P11,P12
rvg
pag
early L1
ventral cord motor neurons
mid L1
B
P cell
Pn.a
Pn.p
hypodermoblast
Neuroblast
W AVF VB X VA VD
P1.a AVF VB X AS VD
P2.a VA VB X AS VD
P3-8.a VA VB VC AS VD
P9-10.a VA VB X AS VD
P11.a VA X X AS VD
P12.a VA X X PDB VD
Legend
A-class
B-class
D-class
ASn
VCn
C
1
2
3
4
5
6
7
8
9
10
Pn.p cell
ventral cord motor neurons
L2

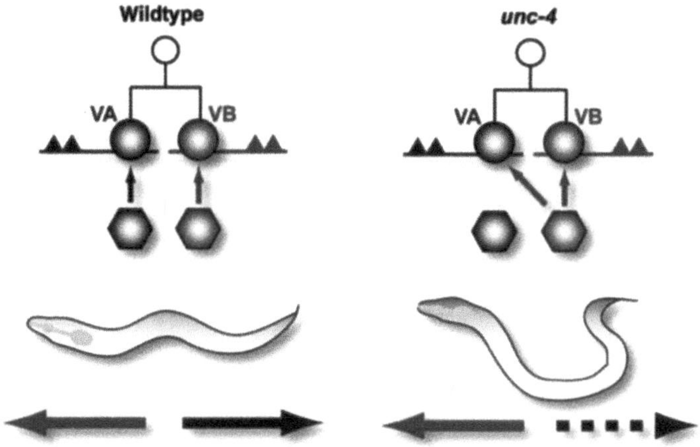

VON STETINA *ET AL.*, FIG. 7. VA motor neurons are miswired with VB-type inputs in *unc-4* mutants. Most VA and VB motor neurons arise from a common precursor and synapse with independent sets of command interneurons to control backward (blue) and forward (red) locomotion, respectively. Inputs to VA motor neurons are from AVA (gap junction, chemical synapse) and from AVD, AVE (chemical synapses). Command interneuron inputs to VBs are from AVB (gap junction) and PVC (chemical synapse). In *unc-4* mutants, VA motor neurons are miswired with inputs normally reserved for VB sisters. As a consequence, *unc-4* mutants are incapable of backward movement (dashed arrow).

VON STETINA *ET AL.*, FIG. 6. Specific blast cells (P-cells) give rise to larval motor neurons in the ventral nerve cord. All animals are viewed from the ventral side. A. L1 larvae. P-cells migrate ventrally from lateral locations in an anterior-to-posterior wave in the early L1. (Adapted from Sulston, 1976). B. The anterior P cell daughter (Pn.a) divides in the late L1 to generate larval motor neurons. Note deviations from stereotypical lineage of P3.a-P8.a for Pn.a cells located at anterior and posterior ends of the VNC (e.g. P1.a). "x" denotes a cell death (adapted from Chalfie and White, 1988). C. L2 larva showing the full complement of ventral cord motor neurons (small circles). Pn.p ectodermal blast cells (numbers) display prominent nucleoli and are reliable landmarks in the queue of ventral cord motor neuron nuclei. (Adapted from Sulston, 1976 and from Sulston and Horvitz, 1977). Additional motor neurons reside in ganglia at either end of the VNC are rvg (retrovesicular ganglion) and pag (preanal ganglion).

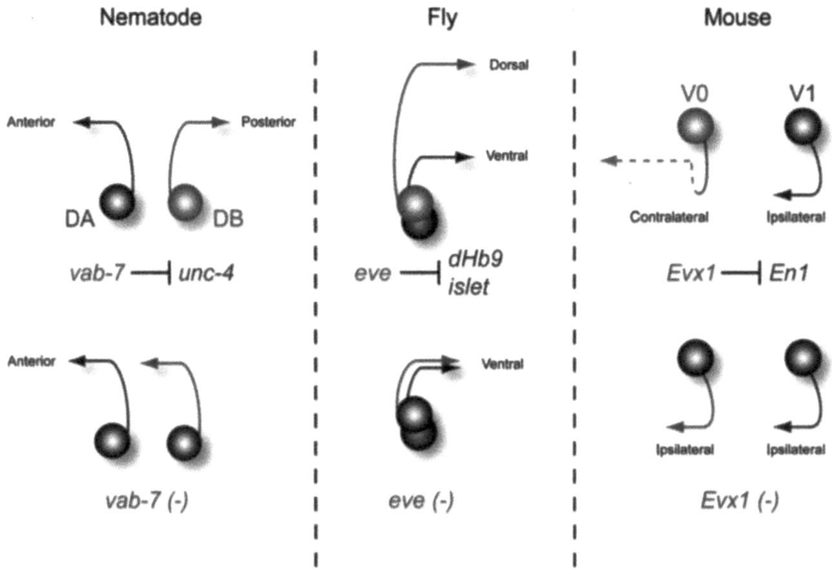

VON STETINA *ET AL.*, FIG. 9. Even-skipped (Eve) transcriptional repressor activity defines neuron identity and axonal trajectory in the axial motor circuits of diverse species. The *C. elegans eve* homolog, *vab-7*, negatively regulates *unc-4* to prevent DB motor neurons from adopting the anterior trajectory of DA motor neurons. In flies, dorsal motor neurons adopt a ventral trajectory when *islet* and *dHB9* are derepressed in *eve* mutants. The mouse *eve* homolog, *Evx1*, represses engrailed to maintain the medial location and contralateral axonal trajectory of V0 interneurons in the spinal cord.

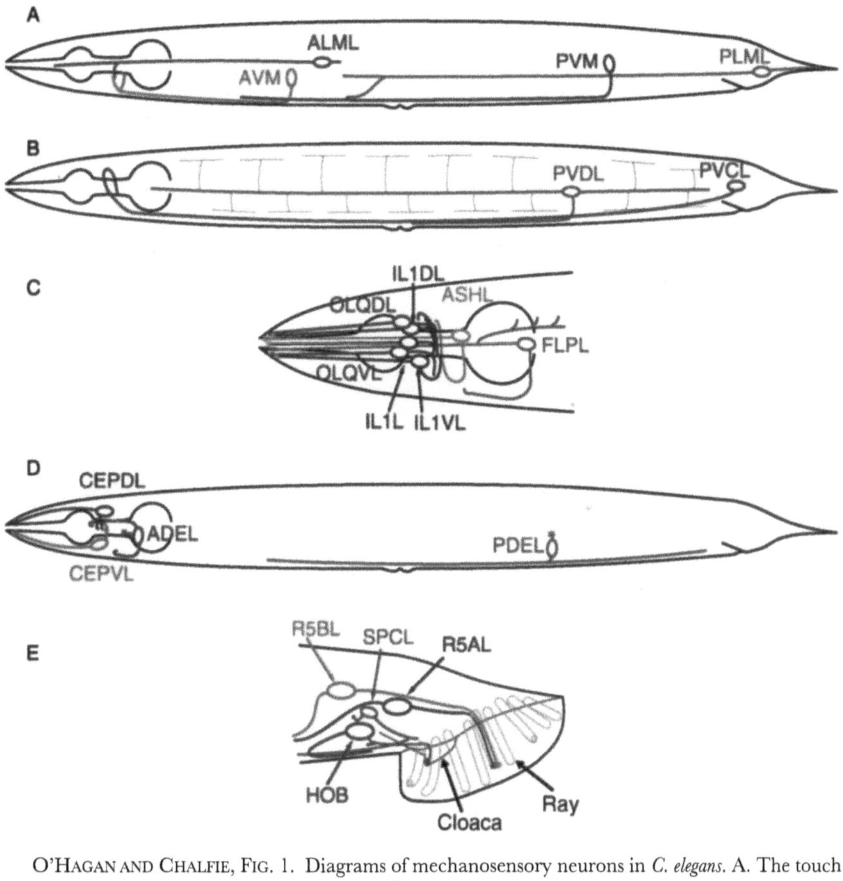

O'HAGAN AND CHALFIE, FIG. 1. Diagrams of mechanosensory neurons in *C. elegans*. A. The touch receptor neurons that sense gentle touch. When a symmetrical pair of cells exists here and in subsequent panels (e.g., ALML and ALMR) only the cell on the left side (ALML) is drawn. B. Neurons needed for harsh touch. C. Neurons that sense nose touch and osmolarity (ASH). Here and in subsequent panels, asterisks mark the positions of sensory cilia. D. Dopaminergic sensory neurons that detect texture. E. Putative mechanosensory cells of the male tail. Panels A to C were redrawn from White *et al.* (1986), panel D was redrawn from White *et al.* (1986) as modified by Tsalik *et al.* (2003), and panel E was redrawn from Emmons and Lipton (2003) and Barr *et al.* (2001).

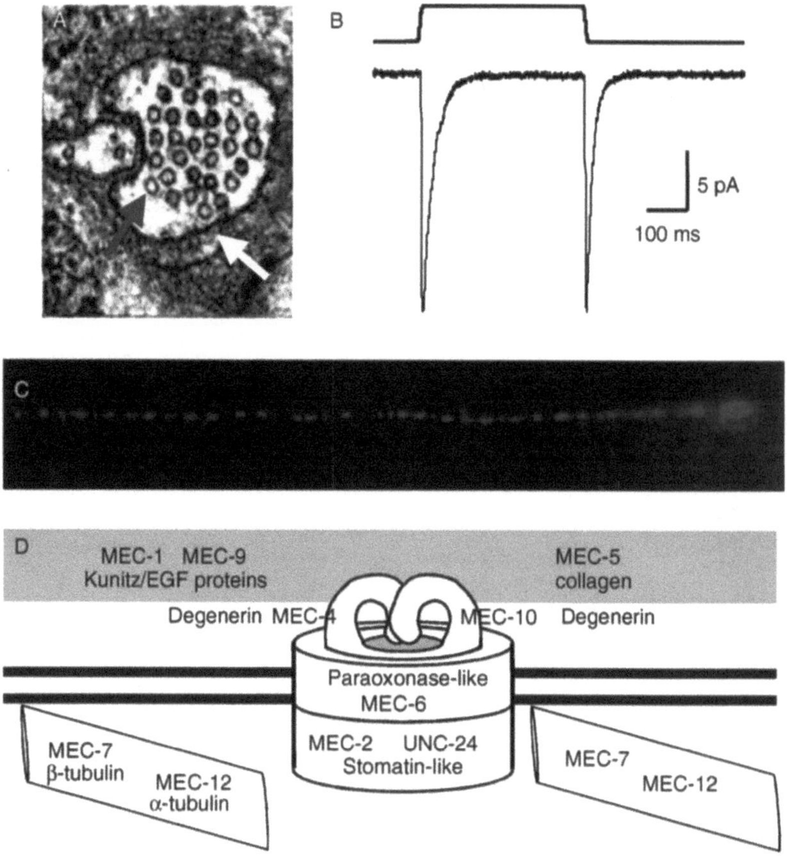

O'HAGAN AND CHALFIE, FIG. 2. Touch receptor proteins and electrophysiological properties. A. Electron micrograph of a touch receptor neuronal process. The process is filled with 15-protofilament microtubules (red arrow) and is surrounded by extracellular matrix (white arrow; modified from Chalfie and Thomson, 1979) B. Wild-type MRC response. The top trace shows the touch stimulus; the bottom trace shows the current response (from O'Hagan *et al.*, 2005). C. Proteins needed for mechanosensation in the touch receptor neurons. D. Punctate distribution of the channel complex as visualized with an antiMEC-2 antibody (from Zhang *et al.*, 2004).

FSC
www.fsc.org
MIX
Papier | Fördert gute Waldnutzung
FSC® C083411

Kolibri 360 GmbH
Ferdinand-Jühlke-Straße 7
99095 Erfurt, Deutschland
produktsicherheit@kolibri360.de